PHYSICS RESEARCH AND TECHNOLOGY

PERSPECTIVES IN MATHEMATICAL PHYSICS

PHYSICS RESEARCH AND TECHNOLOGY

PHYSICS RESEARCH AND TECHNOLOGY

PERSPECTIVES IN MATHEMATICAL PHYSICS

GREGORY D. MONTEGGO
EDITOR

Nova Science Publishers, Inc.
New York

For permission to use material from this book please contact us:
Telephone 631-231-7269; Fax 631-231-8175
Web Site: http://www.novapublishers.com

LIBRARY OF CONGRESS CATALOGING-IN-PUBLICATION DATA

Perspectives in mathematical physics / editor, Gregory D. Monteggo.
p. cm. -- (Physics research and technology)
Includes bibliographical references and index.
ISBN 978-1-61122-984-4 (hardcover : alk. paper)
1. Mathematical physics. I. Monteggo, Gregory D.
QC20.P43 2011
530.15--dc22
2010044742

Published by Nova Science Publishers, Inc. + New York

CONTENTS

PREFACE

This new book presents and discusses current research in the field of mathematical physics. Topics discussed include three dimensional MHD flow between two horizontal plates with transpiration cooling; a solution to the Blasius/Sakiadis boundary layer equation; development of the ignition control algorithm; estafette of drift resonance, stochasticity and particle transport; quantum-magnetic collapse of a neutron gas and modelling of shock waves and clustering in hydrodynamic simulation.

Article 1 – Hydromagnetic effect on the Couette flow between two horizontal parallel porous flat plates with transverse sinusoidal injection of the fluid at the stationary plate and its corresponding removal by constant suction through the plate in uniform motion has been analysed. Due to this type of injection velocity the flow becomes three dimensional. The uniform flow is subjected to a transversely applied magnetic field. The mathematical analysis is presented for the hydromagnetic flow, neglecting the induced magnetic field. Approximate solutions for the velocity field, skin-friction and temperature due to them are obtained and discussed with the help of graphs and tables.

Article 2 – An approximate solution to Blasius/Sakaidis boundary layer equation is derived in terms of error/complimentary error-function. Numerical values of their solution are derived and compared with those of Howarth's (1930) and Sakiadis's (1961) results and are found to agree well. The solution of the thermal boundary layer equation in terms of the complementary error function is also found to agree well when Prandtl number is close to unity.

Article 3 – Local similarity solutions are obtained for the flow of water at 4^0C past a continuously moving semi-infinite plate in the upward direction. Velocity and temperature profiles are shown on graphs and the numerical values of $\{-f''(0)\}$ and $\{-\theta''(0)\}$ are listed in a Table for both Gr/Re^2 (≥ 0) (aiding or opposing flows). The effects of Gr/Re^2 (≥ 0) and viscous dissipative heat on the velocity and temperature field are discussed.

Article 4 – The two dimensional polar fluid flow past an infinite porous plate with constant suction has been considered. Expression for the velocity and the skin friction have been obtained in dimensionless forms exponentially increasing and exponentially decreasing cases. It has been observed that the velocity and the skin friction decrease when there is an increase in couple stress parameters.

Article 5 – The origin and early evolution of super strongly magnetized neutron star pulsars with magnetic fields on the order of 10^{14-15} Gauss, the so-called magnetars, has been under focus in the late years. The suggestion that these objects could appear during the initial evolution of a neutron star, as a consequence of the amplification of the star's remnant magnetic field via the dynamo mechanism, has stimulated an intense research activity. Observations of the Soft Gamma-ray Repeaters (SGRs), a subclass of astrophysical sources emitting powerful 'soft' gamma-ray bursts; with mean energies of 10^{40-44} erg s and $3 - 10$ years time recurrence, have led to the claim that the magnetars may have been discovered. Meanwhile, to put those theoretically classical objects on realistic astrophysical grounds is needed to incorporate to their physical description also the quantum effects. This could be done in the first approximation by superimposing them to the star gravity. In this paper the authors study a model of a neutron star envisioned as a relativistic degenerate neutron gas in equilibrium with a background of electrons and protons acted upon by a magnetic field. It is shown that this system exerts its pressure anisotropically, having a smaller value perpendicular to, instead of along, the magnetic field. For critical fields, as those said typical of magnetars, the magnetic pressure may produce the vanishing of the transversal (to the magnetic field) pressure of the neutron gas, because of its ferromagnetic behavior. Taking it as a model of a neutron star, the outcome could be a transverse collapse of the star. This quantum effect imposes a limit to the fields to be observable in stable neutron star pulsars as a function of their density. The final structure left over after the implosion might be a mixed phase of nucleons and meson condensate, a strange star, or a black string, but no a magnetar at all.

Article 6 – The authors review the development of a control algorithm for ignition burn for an inductive operation in a tokamak reactor including International Thermonuclear Experimental Reactor (ITER). The model based control algorithm for ignition burn is described and its control capability is demonstrated using 0-dimentional simulations. As the basic control algorithm reviewed here is based on the combination of the H-mode power threshold and the fusion power, this system has two inputs and two outputs for feedback. The external heating power is determined by the H-mode power threshold and the fueling rate is by the fusion power. For disturbances such as the change in the confinement factor including L to H and H to L transitions, helium ash particle confinement time, and impurity content, this algorithm sufficiently controls all the burn phases. The developed control algorithm, keeping the beta and the density below their limits, has a fail-safe function in the event of diagnostic failure. In addition, the D-T fuel ratio feedback and pellet injection algorithm are implemented. The diagnostic systems for ignited operation are selected and the diagnostic network system is proposed to back up the control signals for avoiding the control malfunction.

Article 7 – #1.What is the Estafette of Resonances? The question can arise: what is the "estafette,"what does it mean. This word had come from French to Russian and it is known from sport. If one runner with the stick in his hand starts to run and then some time later when one part of the path is already passed brings this stick to another runner, and this runner in his turn will bring the same stick to the following runner the authors say that this is ***estafette***. Such picture can help us to explain what is the estafette in physics, particularly in the physics of a particle transport. In this case the test particle is the "stick" and the adjacent drift resonances are the "runners." If such adjacent resonances are spread in the space, for example, along the radial direction of plasma cross-section, then the particle can be brought across the plasma - from the center to the periphery or vise versa i.e. from the periphery to center. How can the authors arrange and control these resonances. At first this terminology and this approach was proposed in paper, where the removal of the cold alpha-particles in fusion reactor was considered.

2. What is Drift Resonance? Drift resonance is that situation when there is the drift surface with the rational value of the drift rotational transform iota* = m/n, where "m" and "n" are the numbers, and there is the perturbation of the magnetic field with the "wave numbers" "m" and "n". In that case on the place of the rational surface the drift island occurs. Drift islands can take place on the adjacent drift surfaces if they have the rational values of drift rotational transform iota*=m'/n', iota *=m''/n'' and so on and at the same moment there are the magnetic field perturbations with the corresponding sets of wave numbers (m' and n', m'' and n'' and so on).

3. The overlap of adjacent drift resonances can lead to stochasticity if the amplitudes of the perturbation filed are rather large and the distance between the adjacent resonances is rather small.

4. The principal NEW thing here is that perturbations of the magnetic field are being "switched on" consequently one after another. Namely this thing can lead to estafette of resonances and particle transport across the magnetic surface.

5. Now we know that the estafette of drift resonances can take place in the plasma with the ambipolar electric field taken into account. It means that more wide physics phenomenon, particularly Alfven type perturbation and particle interaction in plasma can be understood and described with the use of estafette of resonance.

6. Practical usefulness of the estafette of drift resonances is the removal of cold alpha particles and impurity ions from the plasma core. Applicability can be realized with the use of cooper coils on the as there are on Large Helical Device. Another very important applicability is that estafette of resonances can be the physics mechanism for the interaction of MHD magnetic field perturbations & particles which can lead to the losses of the energetic particles from plasma.

Article 8 – Magnetic reconnection is a topological restructuring of magnetic field lines, leading to a conversion of magnetic energy into kinetic energy and a heating of the plasma. It takes place in regions with strong magnetic gradients: the Earth's magnetopause and the plasma sheet in the magnetotail, during solar and stellar flare events, and in laboratory plasmas on the Earth. It is also involved in astrophysical phenomena like the generation of relativistic jets and processes in accretion disks. The Earth's magnetosphere is the only region where signatures of magnetic reconnection can be measured in-situ by satellites. Based on the method developed by Semenov et al., the authors solve the direct problem of magnetic reconnection and calculate the magnetic field and plasma flow disturbances in a compressible plasma caused by transient reconnection. In order to get the reconnection rate from measured data, we invert the problem. Since the inverse problem is ill–posed, it is treated with the method of regularization. Application of this method to Cluster measurements from September 8_{th}, 2002, where a series of Earth-ward propagating 1–minute scale magnetic field and plasma flow variations were observed outside of the plasma sheet, showed good agreement for the z-component of the reconstructed magnetic field. The reconnection rate is about 1 mV/m and the reconnection region is located at about 24–25 Earth radii in the magnetotail.

Article 9 – The authors show that 2D clustering with constant restitution coefficient is qualitatively well described by a compressible Navier-Stokes like system. Shock capturing highly accurate methods are needed to resolve well shock waves in the cluster formation phase. Numerical results agree with qualitative predictions on clustering.

Versions of these chapters were also published in *Journal of Magnetohydrodyamics, Plasma and Space Research,* Volume 12, Numbers 1-4, and Volume 13, Number 1-4, edited by Frank Columbus, published by Nova Science Publishers, Inc. They were submitted for appropriate modifications in an effort to encourage wider dissemination of research.

In: Perspectives in Mathematical Physics
Editor: Gregory D. Monteggo

ISBN: 978-1-61122-984-4

THREE DIMENSIONAL MHD FLOW BETWEEN TWO HORIZONTAL PLATES WITH TRANSPIRATION COOLING

R.C. Chaudhary**, *Pawan Kumar Sharma*† and *A.K. Mathur
Department of Mathematics, University of Rajasthan, Jaipur-302004, India

ABSTRACT

Hydromagnetic effect on the Couette flow between two horizontal parallel porous flat plates with transverse sinusoidal injection of the fluid at the stationary plate and its corresponding removal by constant suction through the plate in uniform motion has been analysed. Due to this type of injection velocity the flow becomes three dimensional. The uniform flow is subjected to a transversely applied magnetic field. The mathematical analysis is presented for the hydromagnetic flow, neglecting the induced magnetic field. Approximate solutions for the velocity field, skin-friction and temperature due to them are obtained and discussed with the help of graphs and tables.

Keywords: MHD, Couette flow, three-dimensional, transpiration cooling, transverse sinusoidal injection/suction.

1. INTRODUCTION

Transpiration cooling is a very effective process to protact certain structural element in turbojet and rocket engine, like combustion chamber walls, exhaust nozzles, or gas turbine blades from the influence of hot gases. In view of this Eckert [1] obtained an exact solution of the plane Couette flow with transpiration cooling. The problem remained two dimensional due to the uniform injection and suction applied at the porous plates. Flow and heat transfer along a plane wall with periodic suction velocity has been studied by Gersten and Gross [2]. Effects of such a suction velocity on various flow and heat transfer problems along flat and vertical porous plates have been studied by Singh et al [3 ,4] and Singh [5]. Recently the problem of transpiration cooling with the application of the transverse sinusoidal injection/suction velocity has been studied by Singh [6].

* E-mail address: ramacharanchaudhary@rediffmail.com
† E-mail address: pawkumar20@yahoo.co.in

On the other hand in view of the increasing technical application using magnetohydrodynamics effect, it is desirable to extend many of the available viscous hydrodynamic solution to include the effects of magnetic field for those cases when the viscous fluid is electrically conducting. Rossow [7], Greenspan and Carrier [8] and Singh [9,10] have studied extensively the hydromagnetic effects on the flow past a plate with or without injection/suction. The object of the present paper is to study the effects of the magnetic field on the Couette flow between two horizontal parallel porous plates when a transverse sinusoidal injection velocity at the stationary porous plate of the channel is proposed because in actual practice injection or suction may not always be uniform.

2. FOMULATION OF THE PROBLEM

We formulate the governing equations with the assumption that the flow is steady and laminar of a finitely conducting fluid. The magnetic field is applied to the perpendicular to the free stream velocity, at lower magnetic Reynold's number the magnetic field is practically independent of the flow motion and the induced magnetic field is neglected. Conductivity of the fluid is not very large. Further since no external electric field is applied and the effect of polarization of ionized fluid is negligible, hence it can be assumed that the electric field is zero. Therefore under there assumption, we write continuity and momentum equation in the presence of magnetic field as:

$$\nabla . V = 0 \tag{1}$$

$$(V . \nabla) V = - \nabla p / \rho + \nu \nabla^2 V + (J \times B)/\rho \tag{2}$$

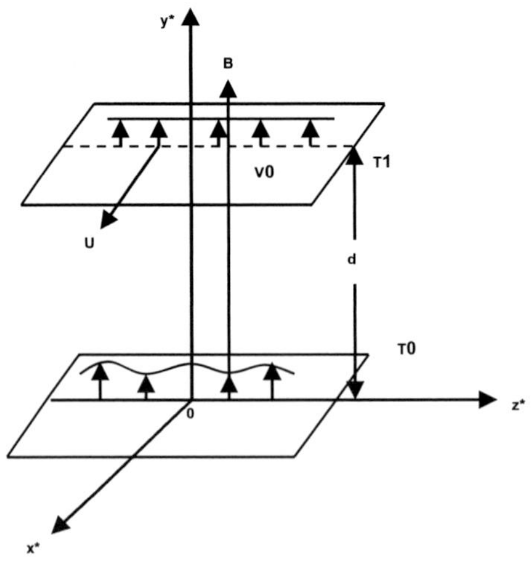

Figure 1. Couette flow with periodic injection and constant suction at the porous plates.

The Lorentz force in equation (2) due to the magnetic field B , modified in the absence of the induced magnetic field is given by

$$(J \times B) = \sigma [(V - V_1) \times B] \times B \tag{3}$$

where V_1 is any suitably chosen characteristic velocity. We now consider the Couette flow of a viscous incompressible fluid between two parallel flat porous plates is considered as shown in Fig.1. The upper plate in uniform motion U is subjected to a constant suction V_0 and the lower to a transverse sinusoidal injection velocity distribution of the form

$$V^*(z^*) = Vo\,[1+\varepsilon \cos(\pi z^*/d)] \tag{4}$$

where ε is positive constant quantity (<<1).Without any loss of generality, the distance d between the plates is taken equal to the wave length of the injection velocity. The lower and upper plates are assumed to be at constant temperature T_0 and T_1 , respectively, with $T_1 > T_0$. All physical quantities are independent of x* for this problem of fully developed laminar flow but the flow remains three-dimensional due to the injection velocity (4). Denoting the velocity component u,v,w in the x, y, z directions, respectively, and the temperature by θ. A magnetic field of uniform strength B is applied perpendicular to the free stream and is along y*-axis. The problem is governed by the following non-dimensional equations

$$v_y + w_z = 0 \tag{5}$$

$$vu_y + wu_z = [u_{yy} + u_{zz} - M^2 (u-1)] / \lambda \tag{6}$$

$$vv_y + wv_z = - p_y + [v_{yy} + v_{zz}] / \lambda \tag{7}$$

$$vw_y + ww_z = - p_z + [w_{yy} + w_{zz} - M^2 w] / \lambda \tag{8}$$

$$v\theta_y + w\theta_z = (\theta_{yy} + \theta_{zz}) / \lambda\, Pr \tag{9}$$

where $y = y^*/ d$, $z = z^*/ d$, $u = u^*/U$, $v = v^*/Vo$, $w = w^*/Vo$, $p = p^*/\rho\, Vo^2$,

Pr (Prandtl number)= ν / α , M (Hartmann number)=$(\sigma B^2 d^2/\rho\nu)^{1/2}$,

λ(injection parameter)=$V_0 d / \nu$, $\theta = (T^* - T_0) / (T_1 - T_0)$

are the dimensionless quantities and ρ ,ν,α,σ and p are respectively, density, kinematic viscosity ,thermal diffusivity, electrical conductivity and pressure. In energy equation a term corresponding to viscous dissipation and Joule heating have been neglected due to their small magnitude as compared to other terms. The (*) stands for dimensional quantities. The corresponding boundary condition in the dimensionless form are :

$$\left.\begin{array}{l} y=0: u=0,\ v(z)=1+\varepsilon\cos\pi z,\ w=0,\ \theta=0, \\ y=1: u=1,\ v=1,\ w=0,\ \theta=1 \end{array}\right\} \quad (10)$$

3. Solution of the Problem

Since the amplitude of the injection velocity ε (<<1) is very small, we now assume the solution of the following form:

$$f(y,z)=f_o(y)+\varepsilon f_1(y,z)+\varepsilon^2 f_2(y,z)+. \quad (11)$$

where f stands for any of u, v, w, p and θ .When ε=0 the problem is reduced to the well known two-dimensional flow with constant injection and suction at both plates. The solution of this two dimensional problem is:

$$u_0(y)=1+[\exp(J_2+J_1y)-\exp(J_1+J_2y)]/[\exp(J_1)-\exp(J_2)] \quad (12)$$

$$\theta_o(y)=[\exp(\lambda Pry)-1]/[\exp(\lambda Pr)-1],\ v_0=1,\ w_0=0,\ p_0=\text{constant}$$

where

$$J_1=[\lambda+(\lambda^2+4M^2)^{1/2}]/2,\ J_2=[\lambda-(\lambda^2+4M^2)^{1/2}]/2$$

When $\varepsilon\neq 0$ substituting (11) in equation (5) to (9) and comparing the coefficient of ε, neglecting those of ε^2, ε^3 etc., the following first order equations are obtained with the help of solutions (12),

$$v_{1y}+w_{1z}=0 \quad (13)$$

$$v_1u_{oy}+u_{1y}=(u_{1yy}+u_{1zz}-M^2u_1)/\lambda \quad (14)$$

$$v_{1y}=-p_{1y}+(v_{1yy}+v_{1zz})/\lambda \quad (15)$$

$$w_{1y}=-p_{1z}+(w_{1yy}+w_{1zz}-M^2w_1)/\lambda \quad (16)$$

$$v_1\theta_{oy}+\theta_{1y}=(\theta_{1yy}+\theta_{1zz})/\lambda\, Pr \quad (17)$$

The corresponding boundary conditions reduce to

$$\left.\begin{array}{llll} y=0: & u_1=0, & v_1=\cos\pi z, & w_1=0,\ \theta_1=0 \\ y=1: & u_1=0, & v_1=0, & w_1=0,\ p_1=0,\ \theta_1=0 \end{array}\right\} \quad (18)$$

This is the set of linear partial differential equations, which describe the three dimensional flow. To solve these equations, we assume v_1, w_1, p_1, u_1 and θ_1 of the following form:

$$\left.\begin{aligned} u_1(y,z) &= u_{11}(y)\cos \pi z \\ v_1(y,z) &= v_{11}(y)\cos \pi z \\ w_1(y,z) &= -\{v'_{11}(y)\sin \pi z\}/\pi \\ p_1(y,z) &= p_{11}(y)\cos \pi z \\ \theta_1(y,z) &= \theta_{11}(y)\cos \pi z \end{aligned}\right\} \quad (19)$$

where a prime denotes differentiation with respect to y. Expressions for $v_1(y,z)$ and $w_1(y,z)$ have been chosen so that the equation of continuity is satisfied. Substituting (19) in (14) to (17), and applying the corresponding boundary conditions, we get the solution for v_1, w_1, p_1, u_1 and θ_1 as:

$$v_1(y,z) = [\, A_1 \exp(J_3 y) + A_2 \exp(J_4 y) - A_3 \exp(J_5 y) - A_4 \exp(J_6 y)\,] \cos \pi z / A. \quad (20)$$

$$w_1(y,z) = -[A_1J_3 \exp(J_3 y) + A_2J_4 \exp(J_4 y) - A_3J_5 \exp(J_5 y) - A_4J_6 \exp(J_6 y)] \sin \pi z / \pi A \quad (21)$$

$$\begin{aligned} p_1(y,z) = [\, & A_1J_3 (J_1 J_3 - \lambda J_3 - M^2) \exp(J_3 y) + A_2J_4 (J_1 J_4 - \lambda J_4 - M^2) \exp(J_4 y) - \\ & -A_3J_5 (J_2J_5 - \lambda J_5 - M^2) \exp(J_5 y) - A_4J_6 (J_2J_6 - \lambda J_6 - M^2) \exp(J_6 y)] \cos\pi z / \pi^2\lambda A \end{aligned} \quad (22)$$

$$\begin{aligned} u_1(y,z) = [& E \exp(g_1y) + F\exp(g_2y) + b_1\{A_1(J_1 \exp(J_2+J_1y+J_3y)/b_2 - J_2 \exp(J_1+J_2y+J_3y)/b_6) \\ & + A_2(J_1\exp(J_2+J_1y+J_4y)/b_3 - J_2\exp(J_1+J_2y+J_4y)/b_7) - A_3(J_1\exp(J_2+J_1y+J_5y)/b_8 \\ & - J_2 \exp(J_1+J_2y+J_5y)/b_4) - A_4(J_1\exp(J_2+J_1y+J_6y)/b_9 - J_2\exp(J_1+J_2y+J_6y)/b_5)\}] \cos\pi z \end{aligned} \quad (23)$$

$$\begin{aligned} \theta_1(y,z) = [& R \exp(s_1y) + S \exp(s_2y) + C_1\{ A_1 \exp(J_3y + \lambda Pry) / C_2 \\ & + A_2 \exp(J_4y + \lambda Pry) / C_3 - A_3 \exp(J_5y + \lambda Pry) / C_4 - A_4 \exp(J_6y + \lambda Pry) / C_5 \}] \cos\pi z \end{aligned} \quad (24)$$

where

$$\begin{aligned} A = & [J_4 J_5 + J_3 J_6 - J_3 J_5 - J_4 J_6] \{\exp(J_5 + J_6) + \exp(J_3 + J_4)\} - [J_4 J_5 + J_3 J_6 - J_3 J_4 - J_5 J_6] \\ & \{\exp(J_3 + J_5) + \exp(J_4 + J_6)\} - [J_5 J_6 + J_3 J_4 - J_3 J_5 - J_4 J_6] \{\exp(J_3 + J_6) + \exp J_4 + J_5)\}, \end{aligned}$$

$$A_1 = [J_4 J_6 - J_5 J_6] \exp(J_4 + J_5) - [J_4 J_5 - J_5 J_6] \exp(J_4 + J_6) + [J_4 J_5 - J_4 J_6] \exp(J_5 + J_6),$$

$$A_2 = [J_3 J_5 - J_5 J_6] \exp(J_3 + J_6) - [J_3 J_6 - J_5 J_6] \exp(J_3 + J_5) - [J_3 J_5 - J_3 J_6] \exp(J_5 + J_6),$$

$$A_3 = [J_4 J_6 - J_3 J_6] \exp(J_3 + J_4) + [J_3 J_4 - J_4 J_6] \exp(J_3 + J_6) - [J_3 J_4 - J_3 J_6] \exp(J_4 + J_6),$$

$$A_4 = [J_3 J_5 - J_4 J_5] \exp(J_3 + J_4) + [J_4 J_5 - J_3 J_4] \exp(J_3 + J_5) - [J_3 J_5 - J_3 J_4] \exp(J_4 + J_5),$$

$$J_3 = [J_1 + (J_1^2 + 4\pi^2)^{1/2}] / 2,$$

$$J_4=[\,J_1-(J_1^2+4\,\pi^2)^{1/2}\,]/2,$$

$$J_5=[\,J_2+(J_2^2+4\,\pi^2)^{1/2}\,]/2,$$

$$J_6=[\,J_2-(J_2^2+4\,\pi^2)^{1/2}\,]/2,$$

$$g_1=[\lambda+(\lambda^2+4\,\pi^2+4\,M^2)^{1/2}\,]/2,\ g_2=[\lambda-(\lambda^2+4\,\pi^2+4\,M^2)^{1/2}\,]/2,$$

$$s_1=[\lambda Pr+(\lambda^2Pr^2+4\,\pi^2)^{1/2}\,]/2,\ s_2=[\lambda Pr-(\lambda^2Pr^2+4\,\pi^2)^{1/2}\,]/2,$$

$$b_1=\lambda/[A\,\{\exp(J_1)-\exp(J_2)\}],\ b_2=3\,J_1\,J_3-\lambda\,J_3,\ b_3=3\,J_1\,J_4-\lambda\,J_4,$$
$$b_4=3\,J_2\,J_5-\lambda\,J_5,\ b_5=3\,J_2\,J_6-\lambda\,J_6,\ b_6=2\,J_2\,J_3+J_1\,J_3-\lambda\,J_3,$$
$$b_7=2\,J_2\,J_4+J_1\,J_4-\lambda\,J_4,\ b_8=2\,J_1\,J_5+J_2\,J_5-\lambda\,J_5,\ b_9=2\,J_1\,J_6+J_2\,J_6-\lambda\,J_6,$$

$$C_1=\lambda^2Pr^2/[A\{\exp(\lambda Pr)-1\}],C_2=J_1J_3+J_3\lambda Pr,\ C_3=J_1J_4+J_4\lambda Pr,$$
$$C_4=J_2J_5+J_5\lambda Pr,\ C_5=J_2J_6+J_6\lambda Pr,C_6=\lambda/[A\{\exp(J_1)-\exp(J_2)\}.\{\exp(g_1)-\exp(g_2)\}],$$
$$C_7=\lambda^2Pr^2/[A\{\exp(\lambda Pr-1\}.\{\exp(s_1)-\exp(s_2)\}],$$

$$E=C_6\,[\,A_1\{J_1(\exp(J_2+g_2)-\exp(J_1+J_2+J_3))/b_2-J_2(\exp(J_1+g_2)-\exp(J_1+J_2+J_3))/b_6\}$$
$$+A_2\{J_1(\exp(J_2+g_2)-\exp(J_1+J_2+J_4))/b_3-J_2(\exp(J_1+g_2)-\exp(J_1+J_2+J_4))/b_7\}$$
$$-A_3\{J_1(\exp(J_2+g_2)-\exp(J_1+J_2+J_5))/b_8-J_2(\exp(J_1+g_2)-\exp(J_1+J_2+J_5))/b_4\}$$
$$-A_4\{J_1(\exp(J_2+g_2)-\exp(J_1+J_2+J_6))/b_9-J_2(\exp(J_1+g_2)-\exp(J_1+J_2+J_6))/b_5\}],$$

$$F=C_6\,[\,A_1\{J_1(\exp(J_1+J_2+J_3)-\exp(J_2+g_1))/b_2-J_2(\exp(J_1+J_2+J_3)-\exp(J_1+g_1))/b_6\}$$
$$+A_2\{J_1(\exp(J_1+J_2+J_4)-\exp(J_2+g_1))/b_3-J_2(\exp(J_1+J_2+J_4)-\exp(J_1+g_1))/b_7\}$$
$$-A_3\{J_1(\exp(J_1+J_2+J_5)-\exp(J_2+g_1))/b_8-J_2(\exp(J_1+J_2+J_5)-\exp(J_1+g_1))/b_4\}$$
$$-A_4\{J_1(\exp(J_1+J_2+J_6)-\exp(J_2+g_1))/b_9-J_2(\exp(J_1+J_2+J_6)-\exp(J_1+g_1))/b_5\}],$$

$$R=C_7\,[A_1\{\exp(s_2)-\exp(J_3+\lambda Pr)\}/C_2+A_2\{\exp(s_2)-\exp(J_4+\lambda Pr)\}/C_3$$
$$-A_3\{\exp(s_2)-\exp(J_5+\lambda Pr)\}/C_4-A_4\{\exp(s_2)-\exp(J_6+\lambda Pr)\}/C_5],$$

$$S=C_7\,[A_1\{\exp(J_3+\lambda Pr)-\exp(s_1)\}/C_2+A_2\{\exp(J_4+\lambda Pr)-\exp(s_1)\}/C_3$$
$$-A_3\{\exp(J_5+\lambda Pr)-\exp(s_1)\}/C_4-A_4\{\exp(J_6+\lambda Pr)-\exp(s_1)\}/C_5].$$

Now after knowing the velocity field, we can calculate skin-friction components τ_x and τ_z in the main and transverse directions, respectively as:

$$\tau_x=d\,\tau_x^*/\mu\,U=(\,d\,u_0/dy\,)_{y=0}+\varepsilon\,(d\,u_{11}/dy\,)_{y=0}\cos\pi z \tag{25}$$

$$\tau_x=C_8+\varepsilon\,[\,E\,g_1+F\,g_2+b_1\{\,A_1((J_1^2+J_1\,J_3)\,.\,\exp(J_2)/b_2-(J_2^2+J_2J_3)\,.\,\exp(J_1)/b_6)$$
$$+A_2((J_1^2+J_1\,J_4)\,.\,\exp(J_2)/b_3-(J_2^2+J_2J_4)\,.\,\exp(J_1)/b_7)$$
$$-A_3((J_1^2+J_1\,J_5)\,.\,\exp(J_2)/b_8-(J_2^2+J_2J_5)\,.\,\exp(J_1)/b_4)$$

$$-A_4((J_1^2 + J_1 J_6).\exp(J_2) / b_9 - (J_2^2 + J_2 J_6) . \exp(J_1)/b_5)\}]\cos\pi z \quad (26)$$

$$\tau_z = d\, \tau_z^* / \mu V_0 = \varepsilon\, (\partial w_1 / \partial y)_{y=0} \quad (27)$$

$$\tau_z = -\, \varepsilon\, [\, A_1 J_3^2 + A_2 J_4^2 - A_3 J_5^2 - A_4 J_6^2\,] \sin \pi z / \pi A \quad (28)$$

Where $C_8 = [\, J_1 \exp(J_2) - J_2 \exp(J_1)] / [\, \exp(J_1) - \exp(J_2)]$

From the temperature field , we can obtain the heat transfer coefficient in terms of Nussult Number as:

$$Nu = d\, .q_w^* / [\, k(T_0 - T_1\,)] = (\, d\, \theta_0 / dy\,)_{y=0} + \varepsilon\, (d\, \theta_{11} / dy\,)_{y=0} \cos \pi z \quad (29)$$

$$Nu = C_9 + \varepsilon\, [R\, s_1 + S\, s_2 + C_1\{\, A_1(J_3 + \lambda Pry) / C_2 + A_2\, (J_4 + \lambda Pry) / C_3 - A_3\, (J_5 + \lambda Pry) / C_4 - A_4\, (J_6 + \lambda Pry) / C_5\, \}]\, \cos\pi z \quad (30)$$

where $C_9 = \lambda Pr / [\{\exp(\, \lambda Pr\,) - 1\}]$

4. Discussion

The main flow velocity profiles are presented in Fig.2., we observe that the velocity increases with the increase of M (Hartmann number) and λ both. The cross flow velocity component w_1 is due to the transverse sinusoidal injection velocity distribution applied through the porous plate at rest. The value of this secondary flow component is given in the table 1. It is evident from the table that w_1 decreases with the increase in the injection parameter λ, upto middle of the channel i.e. $y < 0.5$ and thereafter increases with increasing λ. This is due to the fact that there is injection at the stationary plate and suction at the plate in motion which are two exactly opposite processes. Furthermore it is interesting to note that the transverse velocity decreases near the lower plate but increases away from it near the upper plate, owing to greater Hartmann number.

The Fig.3. depicts the variations of the skin-friction component τ_x and τ_z in the main flow and transverse directions, respectively. We observe that τ_x decreases with increasing M and λ both, while τ_z increases with increasing M and goes on decreasing with increasing λ. The table 2 gives the value of the Nu. The heat transfer coefficient Nu decreases with the increase of injection parameter λ. For the case of air (Pr=0.71) the heat transfer coefficient Nu decreases very slowly with an increase of M, while for the case of water (Pr = 7.0), with an increase in M, the Nu decreases for $\lambda < 0.4$ and increases for $\lambda \geq 0.4$. It is clear from the table that heat transfer coefficient is much lower in the case of water (Pr = 7.0) then that of air (Pr = 0.71).

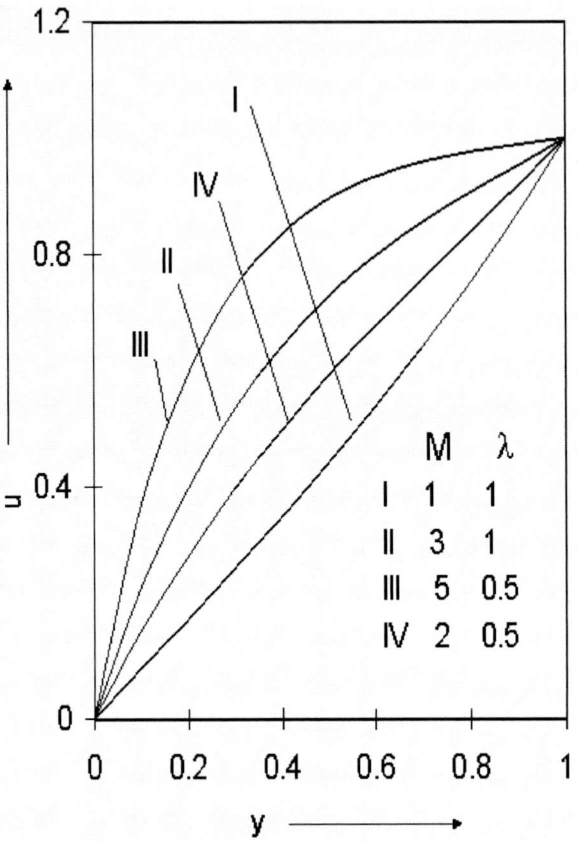

Figure 2. Main flow velocity profiles for z=0 and ε=0.2.

Table 1. The values of w_1 for z = 0.5

y	λ =0.5		λ = 1.0	
	M = 1	M = 3	M = 1	M = 3
0	0	0	0	0
0.1	0.2489	0.2520	0.2409	0.2460
0.2	0.3770	0.3763	0.3686	0.3689
0.3	0.4331	0.4272	0.4263	0.4211
0.4	0.4455	0.4370	0.4417	0.4334
0.5	0.4315	0.4239	0.4314	0.4234
0.6	0.3997	0.3959	0.4033	0.3986
0.7	0.3513	0.3532	0.3581	0.3590
0.8	0.2802	0.2879	0.2886	0.2957
0.9	0.1717	0.1816	0.1789	0.1887
1.0	0	0	0	0

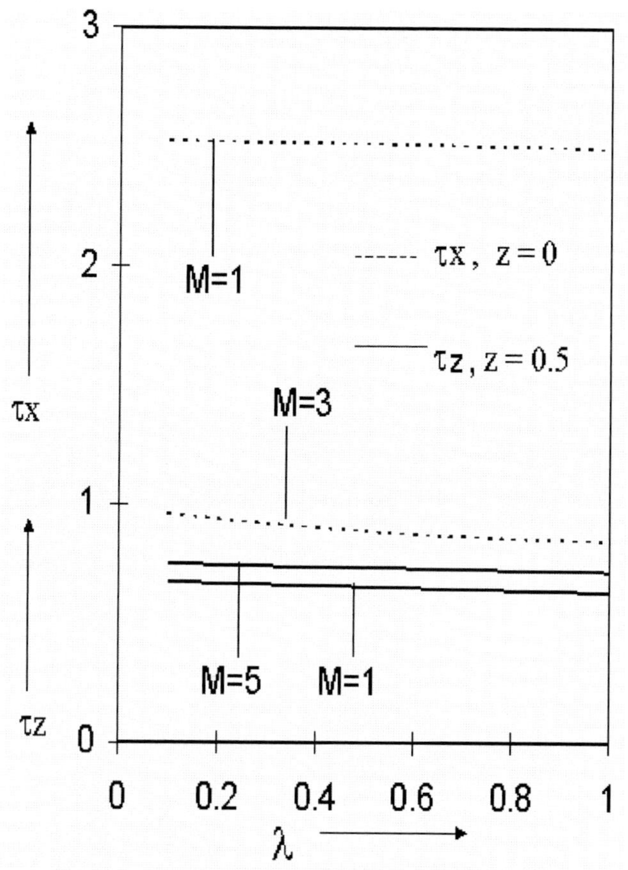

Figure 3. Main flow and transverse components of skin-friction for ε=0.2.

Table 2. The values of the Nusselt number for ε=0.2 and z = 0

λ	Pr = 0.71		Pr =7.0	
	M = 3	M = 5	M = 3	M = 5
0	1	1	1	1
0.1	0.9777	0.9701	0.7605	0.7148
0.2	0.9541	0.9402	0.5388	0.4757
0.3	0.9296	0.9104	0.4111	0.2988
0.4	0.9042	0.8806	0.0873	0.1787
0.5	0.8783	0.8511	0.0875	0.1025
0.6	0.8521	0.8218	0.0522	0.0557
0.7	0.8256	0.7928	0.0289	0.0319
0.8	0.7990	0.7643	0.0154	0.0165
0.9	0.7725	0.7362	0.0079	0.0085
1.0	0.7461	0.7086	0.0040	0.0043

REFERENCES

[1] E. R. Eckert, *Heat and Mass Transfer*, Mcgraw-Hill, New York, 1958.

[2] K.Gersten and J. F. Gross, Flow and heat transfer along a plane wall with periodic suction, *J. Appl. Math. Phys. (ZAMP)* 25 (1974), 399-408.

[3] P. Singh, V. P. Sharma and U. N. Mishra, Three-dimensional free convection flow and heat transfer along a porous vertical wall, *Appl. Sci. Res.* 34 (1978), 105-115.

[4] Ibid, Three-dimensional fluctuating flow and heat transfer along a plate with suction, *Int. J. Heat and Mass Transfer,* 21 (1978), 1117-1123.

[5] K. D. Singh, Three-dimensional viscous flow and heat transfer along a porous plate, *J. Appl. Math. Mech. (ZAMM)* 73 (1993), 58-61.

[6] Ibid, Three-dimensional Couette flow with transpiration cooling, *J. Appl. Math. Phys. (ZAMP)* 50 (1999), 661-668.

[7] V. J. Rossow, On flow of electrically conducting fluids over a flat plate in the presence of a transverse magnetic field, *N. A. C. A., Rept.* 1358, (1958).

[8] H. P. Greenspan and G. F. Carrier, The MHD flow past a flat plate, *J. Fluid Mech.* 6 (1959),
77-96.

[9] K. D. Singh, Hydromagnetic effects on the three-dimensional flow past a porous plate, *J. Appl. Math. Phys. (ZAMP)* 41 (1990), 441-446.

[10] Ibid, Three-dimensional MHD oscillatory flow past a porous plate, *J. Appl. Math. Mech. (ZAMM)* 71 (1991), 192-195.

In: Perspectives in Mathematical Physics
Editor: Gregory D. Monteggo

ISBN: 978-1-61122-984-4

An Approximate Solution to Blasius/Sakiadis Boundary Layer Equation and Corresponding Thermal Boundary Layer Equation

A.K. Singh
Department of Mathematics, Banaras Hindu University, Varanasi- 221 005, India
V.M. Soundalgekar
31A-12, Brindavan Society, Thane-400 601, India
H.S. Takhar
Department of Engineering, Manchester University, Manchester M13 9PL, U.K.

Abstract

An approximate solution to Blasius/Sakaidis boundary layer equation is derived in terms of error/complimentary error-function. Numerical values of our solution are derived and compared with those of Howarth's (1930) and Sakiadis's (1961) results and are found to agree well. The solution of the thermal boundary layer equation in terms of the complementary error function is also found to agree well when Prandtl number is close to unity.

1. Mathematical Analysis

The flow of a viscous incompressible fluid past a semi-infinite horizontal plate was first studied by Blasius (1904) by assuming a series solution and later on accurate numerical results were derived by Howarth (1930). The corresponding flow past a continuously moving semi-infinite plate in a stationary viscous fluid was studied by Sakiadis (1961). The x-axis is taken along the plate in the direction of the flow and the y-axis is taken to the plate. Then with the following similarity transformations,

$$\eta = y\sqrt{U/\upsilon x}, \qquad u = Uf'(\eta), \qquad v = \frac{1}{2}\sqrt{\frac{\upsilon U}{x}}(\eta f' - f) \quad (1)$$

Blasius derived the ordinary non-linear equation from the partial differential equations of Prandtl's boundary layer equations

$$f''' + \frac{1}{2} ff'' = 0 \quad (2)$$

with following boundary conditions

$$f(0) = 0, \qquad f'(0) = 0, \qquad f'(\infty) = 1 \quad (3)$$

Here U is the free-stream velocity and ν is the kinematics viscosity.

The equation governing the flow past a continuously moving plate with velocity U in a stationary fluid is same as in (1) with the following boundary conditions (Sakiadis problem)

$$f(0) = 0, \qquad f'(0) = 1, \qquad f'(\infty) = 0 \quad (4)$$

From the property of error function, we know that

$$erf(0) = 0, \qquad erf(\infty) = 1 \quad (5)$$

and hence we can assume the solution of equation (2) satisfying the boundary conditions (3) in the following form-

$$f'(\eta) = erf(A\eta) \quad (6)$$

where A is a constant to be determined later.

From equation (6), we have

$$f(\eta) = \eta\, erf(A\eta) - \frac{1}{A\sqrt{\pi}}\left(1 - e^{-A^2\eta^2}\right) \quad (7)$$

$$f''(\eta) = \frac{2A}{\sqrt{\pi}} e^{-A^2\eta^2} \quad (8)$$

To determine A, we integrate equation (2) from zero to infinity which then gives

$$\left[f'' + \frac{1}{2}\left(ff' - \int f'^{\,2}\, d\eta\right) \right]_0^\infty = 0 \quad (9)$$

Substituting from equations (6) - (8) into equation (9) and simplifying, we get

$$A^2 = \frac{\sqrt{2} - 1}{4} \quad (10)$$

we have substituted the values of A into equations (5) - (7) and computed the numerical values of f', f'' and these are listed in Table I.

Table I. Comparison of computed values with Howarth (1938)

	$f'(\eta)$		$f''(\eta)$	
	Our results	**Howarth**	**Our results**	**Howarth**
0.0	0.00000	0.00000	0.36311	0.33206
0.2	0.07252	0.06641	0.36161	0.33199
0.4	0.14445	0.13277	0.35714	0.33147
0.6	0.21519	0.19894	0.34982	0.33008
0.8	0.28420	0.26471	0.33982	0.32739
1.0	0.35096	0.32979	0.32739	0.32301
1.2	0.41501	0.39378	0.31281	0.31659
1.4	0.47596	0.45627	0.29641	0.30787
1.6	0.53347	0.51676	0.27855	0.29667
1.8	0.58731	0.57477	0.25961	0.28293
2.8	0.63727	0.62977	0.23996	0.26675
2.2	0.68327	0.68132	0.21997	0.24835
2.4	0.72528	0.72899	0.19998	0.22809
2.6	0.76328	0.77246	0.18031	0.20645
2.8	0.79743	0.81152	0.16123	0.18401
3.0	0.82783	0.84605	0.14298	0.16136
3.2	0.85469	0.87609	0.12575	0.13913
3.4	0.87821	0.90177	0.10969	0.11788
3.6	0.89865	0.92333	0.09488	0.09609
3.8	0.91625	0.94112	0.08140	0.08013
4.0	0.93130	0.95552	0.06926	0.06424
4.2	0.94404	0.96696	0.05844	0.05052
4.4	0.95476	0.97587	0.04891	0.03897
4.6	0.96369	0.98269	0.04059	0.02948
4.8	0.97107	0.98779	0.03341	0.02187
5.0	0.97712	0.99155	0.02727	0.01591
5.2	0.98204	0.33425	0.02208	0.01134
5.4	0.98601	0.99616	0.01773	0.00793
5.6	0.98918	0.99748	0.01412	0.00543
5.8	0.99170	0.99838	0.01115	0.00365
6.0	0.99368	0.99898	0.00873	0.00240
6.2	0.99522	0.99937	0.00678	0.00155
6.4	0.99642	0.99961	0.00522	0.00098
6.6	0.99733	0.99977	0.00399	0.00061
6.8	0.99803	0.99987	0.00302	0.00037
7.0	0.99856	0.99992	0.00227	0.00022
7.2	0.99895	0.99996	0.00169	0.00013
7.4	0.99924	0.99998	0.00125	0.00007
7.6	0.99946	0.99999	0.00092	0.00004

Table I. Continued

	$f'(\eta)$		$f''(\eta)$	
	Our results	**Howarth**	**Our results**	**Howarth**
7.8	0.99961	1.00000	0.00067	0.00002
8.0	0.99973	100000	0.00048	0.00001
8.2	0.99981	1.00000	0.00034	0.00001
8.4	0.99987	1.00000	0.00024	0.00000
8.6	0.99991	1.00000	0.00017	0.00000
8.8	0.99994	1.00000	0.00012	0.00000

We observe from this table that our results agree well with those of Howarth's results, the agreement being excellent near $\eta \sim 1.0$ for $f''(1)$.

Now, Sakiadis (1961) boundary conditions (4) suggests that we assume

$$f'(\eta) = \text{erfc}(B\eta) \tag{11}$$

then

$$f(\eta) = \eta\,\text{erfc}(B\eta) + \frac{1}{B\sqrt{\pi}}(1 - e^{-B^2\eta^2}) \tag{12}$$

$$f''(\eta) = -\frac{2B}{\sqrt{\pi}}\,e^{-B^2\eta^2} \tag{13}$$

Substituting from equations (11) - (13) into equation (9) and simplifying we get

$$B = \frac{\sqrt{2-\sqrt{2}}}{2} \tag{14}$$

We have computed equation (2) along with boundary conditions (3) numerically and also the numerical values of f' and f'' from equations (11) and (13) and these are listed in Table II. We observed that they agree very well.

Table II. Comparision of computed values (Sakiadis problem)

	$f'(\eta)$		$f''(\eta)$	
	Analytically	**Numerically**	**Analytically**	**Numerically**
0.0	1.00000	1.00000	-0.43181	-0.44390
0.2	0.91381	0.91151	-0.43181	-0.43962
0.4	0.82861	0.82468	-0.42181	-0.42751
0.6	0.74539	0.74094	-0.40964	-0.40892
0.8	0.66505	0.66144	-0.39318	-0.38540

Table II. Continued

	$f'(\eta)$		$f''(\eta)$	
	Analytically	**Numerically**	**Analytically**	**Numerically**
1.0	0.58837	0.58701	-0.37299	-0.35844
1.2	0.51606	0.51819	-0.34971	-0.32948
1.4	0.44865	0.45527	-0.32407	-0.29974
1.6	0.38654	0.39828	-0.29681	-0.27020
1.8	0.32998	0.34712	-0.26868	-0.24164
2.0	0.27908	0.30153	-0.24038	-0.21460
2.2	0.23380	0.26115	-0.21255	-0.18944
2.4	0.19399	0.22561	-0.18576	-0.16636
2.6	0.15939	0.19447	-0.16045	-0.14543
2.8	0.12968	0.16730	-0.13698	-0.12664
3.0	0.10446	0.14367	-0.11558	-0.10991
3.2	0.08330	0.12320	-0.09639	-0.09511
3.4	0.06576	0.10551	-0.07945	-0.08211
3.6	0.05138	0.09025	-0.06472	-0.07073
3.8	0.03973	0.07712	-0.05211	-0.06082
4.0	0.03040	0.06584	-0.04146	-0.05221
4.2	0.02302	0.05616	-0.03261	-0.04477
4.4	0.01725	0.04786	-0.02535	-0.03834
4.6	0.01279	0.04076	-0.01948	-0.03281
4.8	0.57105	0.03469	-0.01479	-0.02805
5.0	0.00681	0.02950	-0.01110	-0.02396
5.2	0.00489	0.02506	-0.00823	-0.02046
5.4	0.00347	0.02128	-0.00604	-0.01746
5.6	0.00244	0.01805	-0.00437	-0.01489
5.8	0.00170	0.01530	-0.00313	-0.01270
6.0	0.00117	0.01295	-0.00222	-0.01083
6.2	0.00079	0.01095	-0.00155	-0.00923
6.4	0.00053	0.00924	-0.00107	-0.00786
6.6	0.00035	0.00779	-0.00073	-0.00670
6.8	0.00023	0.00655	-0.00049	-0.00570
7.0	0.00015	0.00550	-0.00033	-0.00486
7.2	0.00010	0.00460	-0.00022	-0.00414
7.4	0.00006	0.00384	-0.00014	-0.00352
7.6	0.00004	0.00319	-0.00009	-0.00300
7.8	0.00002	0.00263	-0.00006	-0.00255
8.0	0.00001	0.00216	-0.00004	-0.00217
8.2	0.00001	0.00176	-0.00002	-0.00185
8.4	0.00001	0.00142	-0.00001	-0.00158
8.6	0.00000	0.00113	-0.00001	-0.00134
8.8	0.00000	0.00088	-0.00001	-0.00114

2. SOLUTION OF THERMAL BOUNDARY LAYER EQUATION

The similarity variable defined in Eq. (1), the variable $\theta(\eta)$ and Prandtl number Pr defined as

$$\theta(\eta) = (T - T_\infty) / (T_w - T_\infty), \qquad Pr = \mu C_p / k \tag{15}$$

leads the following ordinary differential equation

$$\theta'' + \frac{1}{2} Pr\, f\, \theta' = 0 \tag{16}$$

from the partial differential equation of the thermal boundary layer equation. In equation (15), T is the fluid temperature in the thermal boundary layer; T_w, the temperature of the plate; T_∞, the temperature of the fluid away from the plate; μ, the dynamic viscosity; C_p, specific heat at constant pressure and k, the thermal conductivity of the fluid. The boundary conditions corresponding to equation (16) are

$$\theta(0) = 1 \quad \text{and} \quad \theta(\infty) = 0 \tag{17}$$

Table III. Values of $\theta(\eta)$ corresponding to Blasius problem

	Pr = 0.71		Pr = 1.0		Pr = 1.2	
	Analytically	**Numerically**	**Analytically**	**Numerically**	**Analytically**	**Numericall y**
0.0	1.00000	1.00000	1.00000	1.00000	1.00000	1.00000
0.2	0.93887	0.94117	0.92748	0.93359	0.92058	0.92921
0.4	0.87809	0.88236	0.85555	0.86723	0.84194	0.85849
0.6	0.81803	0.82368	0 .78481	0.80105	0.76485	0.78800
0.8	0.75902	0.76524	0 .71580	0.73526	0.69003	0.71802
1.0	0.70137	0.70724	0.64904	0.67018	0.61811	0.64893
1.2	0.64540	0.64993	0.58499	0.60617	0.54969	0.58121
1.4	0.59137	0.59358	0.52404	0.54368	0.48522	0.51539
1.6	0.53952	0.53851	0.46653	0.48317	0.42508	0.45209
1.8	0.49004	0.48506	0.41269	0.42515	0.36953	0.39190
2.0	0.44312	0.43359	0.36273	0.37013	0.31874	0.33542
2.2	0.39888	0.38443	0.31673	0.31857	0.27275	0.28317
2.4	0.35741	0.33793	0.27474	0.27089	0.23152	0.23561
2.6	0.31876	0.29438	0.23672	0.22740	0.19492	0.19302
2.8	0.28296	0.25402	0.20257	0.18834	0.16275	0.15558
3.0	0.24998	0.21704	0.17217	0.15380	0.13476	0.12327
3.2	0.21979	0.18355	0.14531	0.12377	0.11065	0.09595
3.4	0.19231	0.15359	0.12179	0.09809	0.09008	0.07332
3.6	0.16744	0.12712	0.10135	0.07653	0.07270	0.05497
3.8	0.14507	0.10404	0.08375	0.05875	0.05817	0.04041
4.0	0.12506	0.08418	0.06870	0.04436	0.04614	0.02911
4.2	0.10728	0.06731	0.05596	0.03293	0.03628	0.02055
4.4	0.09156	0.05319	0.04524	0.02403	0.02827	0.01420
4.6	0.07774	0.04153	0.03631	0.01723	0.02184	0.00961

Table III. Continued

	Pr = 0.71	Pr = 1.0	Pr = 1.2		Pr = 0.71	Pr = 1.0
	Analytically	**Numerically**	**Analytically**	**Numerically**	**Analytically**	**Numerically**
4.8	0.06568	0.03202	0.02893	0.01214	0.01671	0.00636
5.0	0.05520	0.02439	0.02288	0.00840	0.01268	0.00412
5.2	0.04615	0.01835	0.01796	0.00571	0.00953	0.00261
5.4	0.03839	0.01362	0.01399	0.00381	0.00710	0.00162
5.6	0.03176	0.00999	0.01082	0.00249	0.00524	0.00098
5.8	0.02614	0.00723	0.00830	0.00160	0.00383	0.00058
6.2	0.01743	0.00364	0.00478	0.00063	0.00200	0.00019
6.4	0.01412	0.00254	0.00358	0.00038	0.00142	0.00011
6.6	0.01138	0.00174	0.00267	0.00023	0.00100	0.00006
6.8	0.00912	0.00118	0.00197	0.00013	0.00070	0.00003
7.0	0.00727	0.00079	0.00144	0.00008	0.00048	0.00002
7.2	0.00576	0.00052	0.00105	0.00004	0.00033	0.00001
7.4	0.00454	0.00034	0.00076	0.00002	0.00023	0.00000
7.6	0.00356	0.00022	0.00054	0.00001	0.00015	0.00000
7.8	0.00278	0.00014	0.00039	0.00001	0.00010	0.00000
8.0	0.00216	0.00009	0.00027	0.00000	0.00007	0.00000
8.2	0.00166	0.00005	0.00019	0.00000	0.00004	0.00000
8.4	0.00128	0.00003	0.00013	0.00000	0.00003	0.00000
8.6	0.00097	0.00002	0.00009	0.00000	0.00002	0.00000
8.8	0.00074	0.00001	0.00006	0.00000	0.00001	0.00000

The solution of equation (16) with boundary conditions (17) was first given by Pohlhausen (1921) as

$$\theta(\eta) = \int_{\xi=\eta}^{\infty} \left[f^{//}(\xi)\right]^{\Pr} d\xi \Big/ \int_{\xi=0}^{\infty} [f^{//}(\xi)]^{\Pr} d\xi \qquad (18)$$

In the case of Blasius problem, using the expression for $f^{//}(\xi)$ given in equation (8), we have

$$\theta(\eta) = erfc(\sqrt{\Pr}\, A\eta) \qquad (19)$$

From above equation, the rate of heat transfer at the plate is given by

$$\theta^{/}(0) = -\,2A\sqrt{\Pr/\,\pi} \qquad (20)$$

The use of expression $f^{//}(\xi)$ given by equation (13) results the following solution

$$\theta(\eta) = erfc(\sqrt{\Pr}\, B\eta) \qquad (21)$$

in the case of Sakiadis problem. For this case

$$\theta'(0) = -2B\sqrt{Pr/\pi} \tag{22}$$

Comparison of the numerical values of $\theta(\eta)$ obtained from equation (19) for Pr = 0.71, 1.0 and 1.2 with the values obtained by solving the governing equation by a numerical method given in Table IV clearly indicates that the agreement between them is good. Same results can also be observed from the next table in the case of moving plate. In Table V, the values of $\theta'(0)$ are given for different values of Pr obtained from equations (20) and (22). Comparison of these values to those of numerically obtained values clearly shows that the agreement is good in both cases when Pr ≈ 1.

Table IV. Values of $\theta(\eta)$ corresponding to Sakiadis problem

	Pr = 0.71		**Pr = 1.0**		**Pr = 1.2**	
	Analytically	**Numerically**	**Analytically**	**Numerically**	**Analytically**	**Numericall y**
0.0	1.00000	1.00000	1.00000	1.00000	1.00000	1.00000
0.2	0.92733	0.92931	0.91381	0.91147	0.90562	0.90052
0.4	0.85526	0.85956	0.82861	0.82460	0.81255	0.80329
0.6	0.78438	0.79159	0.74539	0.74083	0.72206	0.71020
0.8	0.71525	0.72609	0.66505	0.66130	0.63530	0.62274
1.0	0.64838	0.66358	0.58837	0.58685	0.55328	0.54194
1.2	0.58422	0.60446	0.51606	0.51805	0.47682	0.46839
1.4	0.52320	0.54898	0.44865	0.45509	0.40654	0.40234
1.6	0.46562	0.49728	0.38654	0.39810	0.34284	0.34371
1.8	0.41174	0.44939	0.32998	0.34694	0.28591	0.29219
2.0	0.36175	0.40526	0.27908	0.30134	0.23574	0.24733
2.2	0.31574	0.36479	0.23380	0.26098	0.19214	0.20856
2.4	0.27376	0.32781	0.19399	0.22544	0.15478	0.17530
2.6	0.23576	0.29415	0.15939	0.19431	0.12322	0.14691
2.8	0.20165	0.26360	0.12968	0.16714	0.09692	0.12282
3.0	0.17129	0.23594	0.10446	0.14353	0.07531	0.10245
3.2	0.14449	0.21096	0.08330	0.12307	0.05781	0.08530
3.4	0.12103	0.18843	0.06576	0.10539	0.04383	0.07091
3.6	0.10066	0.16815	0.05138	0.09014	0.03282	0.05886
3.8	0.08312	0.14993	0.03973	0.07702	0.02427	0.04879
4.0	0.06814	0.13356	0.03040	0.06575	0.01772	0.04041
4.2	0.05546	0.11888	0.02302	0.05607	0.01278	0.03343
4.4	0.04480	0.10573	0.01725	0.04779	0.00909	0.02763
4.6	0.03593	0.09395	0.01279	0.04069	0.00639	0.02283
4.8	0.02860	0.08341	0.00938	0.03463	0.00443	0.01884
5.0	0.02260	0.07398	0.00681	0.02944	0.00303	0.01554
5.2	0.01773	0.06555	0.00489	0.02502	0.00205	0.01281
5.4	0.01380	0.05802	0.00347	0.02124	0.00137	0.01055
5.6	0.01066	0.05129	0.00244	0.01801	0.00090	0.00868
5.8	0.00817	0.04528	0.00170	0.01527	0.00058	0.00714
6.0	0.00622	0.03991	0.00117	0.01292	0.00037	0.00587
6.2	0.00469	0.03512	0.00079	0.01092	0.00024	0.00482

Table IV. Continued

	Pr = 0.71		Pr = 1.0		Pr = 1.2	
	Analytically	Numerically	Analytically	Numerically	Analytically	Numerically
6.4	0.00352	0.03085	0.00053	0.00922	0.00015	0.00395
6.6	0.00261	0.02703	0.00035	0.00777	0.00009	0.00323
6.8	0.00193	0.02362	0.00023	0.00654	0.00006	0.00264
7.0	0.00141	0.02058	0.00015	0.00548	0.00003	0.00216
7.2	0.00103	0.01787	0.00010	0.00459	0.00002	0.00176
7.4	0.00074	0.01545	0.00006	0.00383	0.00001	0.00143
7.6	0.00053	0.01330	0.00004	0.00318	0.00001	0.00115
7.8	0.00038	0.01137	0.00002	0.00262	0.00000	0.00093
8.0	0.00026	0.00965	0.00001	0.00215	0.00000	0.00074
8.2	0.00018	0.00812	0.00001	0.00175	0.00000	0.00059
8.4	0.00013	0.00676	0.00001	0.00141	0.00000	0.00046
8.6	0.00009	0.00554	0.00000	0.00112	0.00000	0.00036
8.8	0.00006	0.00445	0.00000	0.00088	0.00000	0.00027

Table. V. Numerical values of { $-\theta'(0)$} for different values of Prandtl number (Pr)

Pr	For Blasius problem		For Sakiadis problem	
	From Eq. (20)	Numerically	From Eq. (22)	Numerically
0.044	0.07616	0.11129	0.09057	0.11385
0.6	0.28126	0.27697	0.34480	0.31701
0.71	0.30596	0.29418	0.36851	0.35468
0.8	0.32477	0.30694	0.38622	0.38045
1.0	0.36311	0.32085	0.43181	0.44483
1.1	0.38083	0.34337	0.45288	0.47316
1.2	0.39776	0.35397	0.47302	0.50029
7.0	0.96069	0.64622	1.14246	1.40713

REFERENCES

H. Blasius, (1908), Grenzschichten in Flusigkeiken mit Kleiner Reibung, Z. *Math. Phys*., 56, 1-37.

L. Howarth (1938), On the solution of the laminar boundary layer equations, *Proc. Roy. Soc. (London),* A164, 5497-579.

E. Pohlhausen, (1921), Der warmeaustausch zwischen festen korpern und Flussigkeisen mit Kliener Reibung und Kleiner Warmeleitung, *ZAMM* 1, pp. 115-121.

B. C. Sakiadis, (1961), Boundary layer behaviour on continuous solid surfaces : I. Boundary layer equations for two-dimensional and axisymmetric flow, *Amer. Inst. Chem. Eng. J*., 7, 26-28.

In: Perspectives in Mathematical Physics
Editor: Gregory D. Monteggo
ISBN: 978-1-61122-984-4

On Flow of Water at 4^0C Past a Continuously Moving Semi-Infinite Vertical Plate in the Upward Direction

V.M. Soundalgekar
31a-12, Brindavan Society, Thane(400601), India
H.S. Takhar*
Department of Engineering, Manchester Metropolitan University
Manchester, M1 5GD, UK
B.S. Jaiswal
Birla College, Kalyan (421304), India
R.K. Deka
Department of Mathematics
Gauhati University, Guwahati (Assam) (781014), India

Abstract

Local similarity solutions are obtained for the flow of water at 4^0C past a continuously moving semi-infinite plate in the upward direction. Velocity and temperature profiles are shown on graphs and the numerical values of $\{-f''(0)\}$ and $\{-\theta''(0)\}$ are listed in a Table for both Gr/Re^2 (≥ 0) (aiding or opposing flows). The effects of Gr/Re^2 (≥ 0) and viscous dissipative heat on the velocity and temperature field are discussed.

1. Introduction

The flow of a viscous incompressible fluid past a semi-infinite horizontal plate was first studied by Blasius (1908) by considering the boundary layer equations. However, more accurate numerical solution to the velocity and temperature field was derived by Howarth (1936, 1938). Later on, it was shown by Sakiadis (1961) that the flow past a continuously moving semi-infinite plate is different from that of the boundary layer flow studied by Blasius (1908). The heat transfer aspect of the problem studied by Sakiadis (1961) was later on studied by Tsou et al. (1967).

* To whom correspondence should be addressed

The combined free and forced convection flow past a semi-infinite vertical plate of an incompressible viscous fluid was presented by Sparrow et al. (1959a) who presented a similarity solution by assuming both the non-isothermal wall temperature and non-uniform external flow to be specified in a particular manner. Later on, Sparrow and Gregg (1959b) presented a series solution to the combined free and forced convection flow past a semi-infinite vertical plate. The series was in power of Gr/Re2 where Gr and Re are the Grashof number (=gβ(T-T∞)x3/ν2) and the Reynolds number (=xU0/ν) respectively. Acrivos (1958) solved the problem of ref. (7) by using the Karman-Pohlhausen method. The experimental confirmation of these results was carried out by Kliegel (1959), for both the aiding and opposed cases. Szewczyk (1964) again studied this combined free and forced convection problem by expanding the stream-function and temperature function into series in terms of the parameters Gr/Re2up to the second approximation.

However, all these studies are for fluids such as air (Pr = 0.71), water (Pr = 7.0) at normal room temperature of 200C. At this normal temperature, it is well established that the difference between the densities at any two points is given by a linear function of the difference between the temperatures at the same two points and is defined as,

$$\Delta\rho = -\rho\beta(\Delta T) \tag{1.1}$$

where β is the coefficient of thermal expansion. But in the case of water at 4^0C, where the density of water is maximum, equation (1.1) gets modified and is given by,

$$\Delta\rho = -\rho\gamma(\Delta T^2) \tag{1.2}$$

where $\gamma = 8\times10^6 K^{-2}$ is the thermal expansion coefficient. Taking this fact into account, free convection flow of water at 4^0C past a semi-infinite vertical plate was studied by Goren (1966), whereas the combined free and forced convection flow of water at 4^0C past a semi-infinite vertical isothermal plate was studied by Soundalgekar et al. (1984). Now, instead of the vertical plate held stationary and fluid being made to flow past such a vertical semi-infinite plate, following Sakiadis (1961), if the plate is moved continuously in its own plane, through a stationary mass of water at 4^0C, then how is the flow of water at 4^0C affected? This has not been studied in the literature. Only recently, Soundalgekar et al. (1999) studied the case of flow past a continuously moving plate in the vertically upward direction for fluids at normal temperature of 20^0C.

It is now proposed to study the flow of water at 4^0C past a continuously moving plate in the upward direction, without neglecting viscous dissipative heat. In Section 2, the mathematical analysis is presented and in Section 3, the conclusions are mentioned.

MATHEMATICAL ANALYSIS

Consider the two dimensional flow of water at 4^0C past a semi-infinite vertical plate, moving in an upward direction continuously. The x-axis is taken along the plate in the vertically upward direction while the y-axis is taken normal to the plate, with the origin at the leading edge of the plate. Under the usual Boussinesq's approximation, the flow of water at 4^0C can be shown to be governed by the following boundary layer equations:

Continuity:

$$\frac{\partial u}{\partial x}+\frac{\partial v}{\partial y}=0 \tag{2.1}$$

Momentum:

$$u\frac{\partial u}{\partial x}+v\frac{\partial u}{\partial y}=\nu\frac{\partial^2 u}{\partial y^2}+g\gamma(T-T_\infty)^2 \tag{2.2}$$

Energy:

$$u\frac{\partial T}{\partial x}+v\frac{\partial T}{\partial y}=\alpha\frac{\partial^2 T}{\partial y^2}+\frac{\mu}{\rho C_P}\left(\frac{\partial u}{\partial y}\right)^2 \tag{2.3}$$

The boundary conditions are given by,

$$\left.\begin{array}{lll} u=U_0, \quad \text{v}=0, \quad T=T & \text{at} & \text{y}=0 \\ u=0, \quad \text{v}=0, \quad T=T_\infty & \text{as} & \text{y}\to\infty \end{array}\right\} \tag{2.4}$$

Here u, v are the velocity components of the fluid in the x and y-directions respectively, ν is the kinematic viscosity, γ is the thermal expansion coefficient, g the acceleration due to gravity, T the temperature in the boundary layer, T_w the plate temperature, $\alpha(=K/\rho C_P)$ the thermal diffusivity, K the thermal conductivity, ρ the density, C_P the specific heat at constant temperature, U_0 the free stream velocity and T_∞ is the temperature of the free stream.

Equations (2.1) – (2.4) reduce to following ordinary differential equations

$$f'''+\frac{1}{2}ff''+\frac{Gr}{\text{Re}^2}\theta^2=0 \tag{2.5}$$

$$\theta''+\frac{1}{2}\Pr f\theta'+\Pr E f''^2=0 \tag{2.6}$$

with the following boundary conditions:

$$\left.\begin{array}{l} f(0)=0, \quad f'(0)=1, \quad \theta(0)=1 \\ f'(\infty)=0, \quad \theta(\infty)=0 \end{array}\right\} \tag{2.7}$$

by using the following transformations:

$$\left.\begin{aligned} &\eta = y\left(\frac{U_0}{\nu x}\right)^{1/2}, \quad u = U_0 f'(\eta), \quad v = \frac{1}{2}\left(\frac{U_0 \nu}{x}\right)^{1/2}(\eta f' - f), \quad \theta = \frac{T - T_\infty}{T_w - T_\infty} \\ &\Pr = \frac{\mu C_P}{K}, \quad E = \frac{U_0^2}{C_P (T_w - T_\infty)}, \quad Gr = \frac{g\gamma (T_w - T_\infty)^2 x^3}{\nu^2}, \quad \mathrm{Re} = \frac{U_0 x}{\nu} \end{aligned}\right\} \quad (2.8)$$

Here Pr is the Prandtl number of water (11.4 at 4^0C and 7 at 20^0C, E the Eckert number, Gr the Grashof number and Re is the Reynolds number. Equations (2.5) – (2.6) are not exact similarity equations, but are locally non-similar equations.

These are solved numerically on a computer and the results for velocity and temperature profiles are shown on Figures 1 and 2 respectively. We observe from Figure 1 that in the case of aiding flow ($Gr/Re^2 > 0$), the velocity increases with increasing Gr/Re^2, but in the opposing flow, it decreases. The velocity also increases due to more viscous dissipative heat. From Figure 2, we observe that in case of aiding flow, the temperature decreases due to an increase in the value of Gr/Re^2 whereas in opposing flow, it increases. Also due to greater viscous dissipative heat, there is a rise in the temperature of water in aiding flow and in opposing flow, the temperature is not significantly affected by greater viscous dissipative heat.

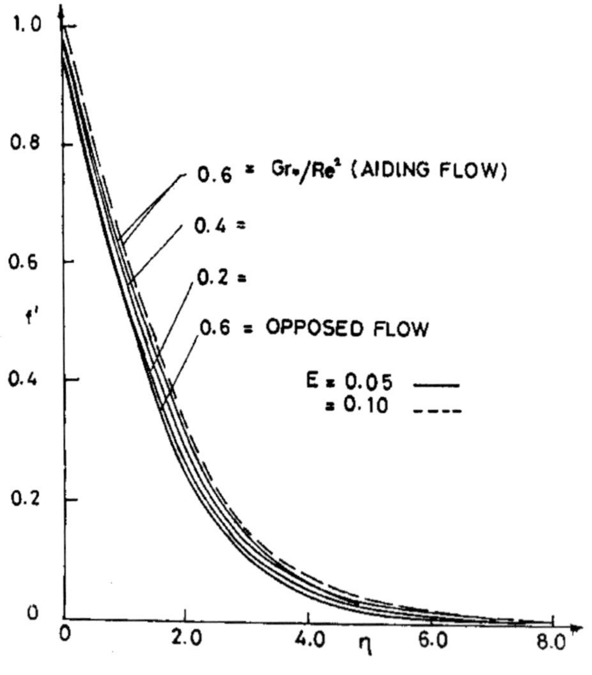

Figure 1. Velocity Profiles. Pr = 11.4

On Figure 2, comparison of velocity profiles for Pr = 11.4 and 7.0 is made. We observe from this Figure 3(a) that when $Gr/Re^2 > 0$, the velocity is higher in the case of water at 4^0C as compared to that at 20^0C whereas for $Gr/Re^2 < 0$, opposite the case. On Figure 3(b), the effect of viscous dissipative heat on the velocity profiles is shown for both Pr = 11.4 and Pr = 7.0 when $Gr/Re^2 > 0$. We observe from this figure that greater viscous dissipative heat causes a rise in the velocity of water at Pr = 7.0 i.e. at a room temperature of 20^0C.

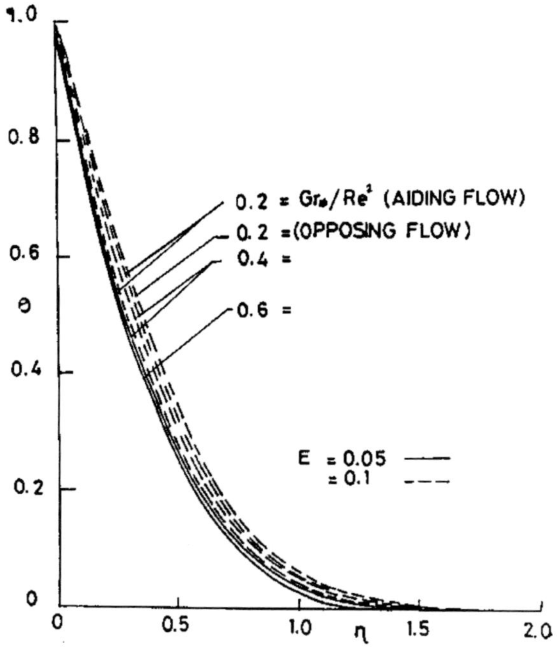

Figure2. Temperature Profiles. Pr = 11.5, E = 0.1.

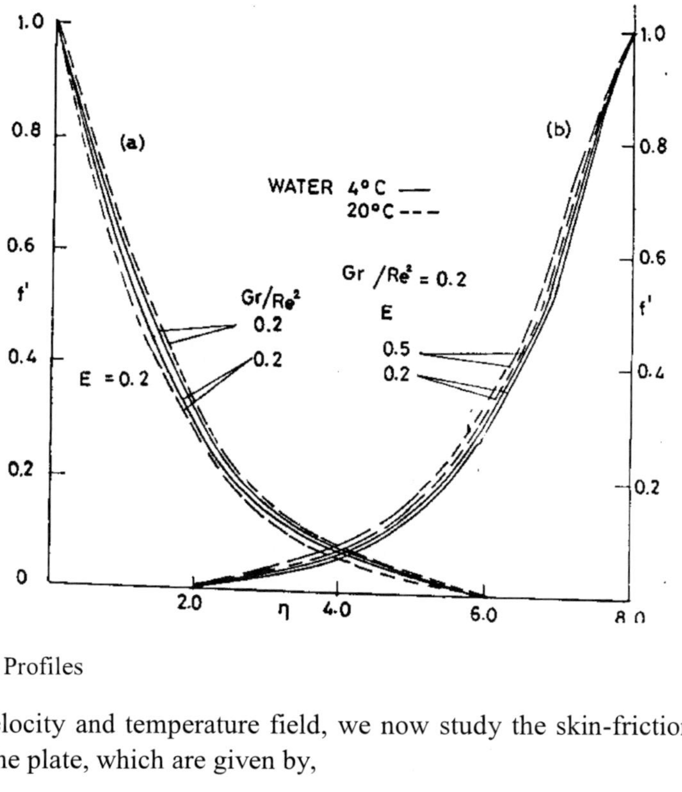

Figure 3. Velocity Profiles

From the velocity and temperature field, we now study the skin-friction and the rate of heat transfer at the plate, which are given by,

$$\tau' = -\mu \frac{\partial u}{\partial y}\bigg|_{y=0} \quad (2.9)$$

$$q = -K \frac{\partial T}{\partial y}\bigg|_{y=0} \quad (2.10)$$

and in view of (2.8), we have

$$\tau = \frac{\tau'}{\rho U_0^2} = -\frac{1}{\sqrt{\mathrm{Re}}} f''(0) \quad (2.11)$$

$$Nu = -\sqrt{\mathrm{Re}}\theta'(0) \quad (2.12)$$

where $Nu = \dfrac{qx}{K(T_w - T_\infty)}$

The numerical values of $\{-f''(0)\}$ and $\{-\theta'(0)\}$ are computed and these are listed in Table 1. We observe from table 1 that in the case of aiding flow, an increase in Gr/Re^2 leads to a decrease in the values of $\{-f''(0)\}$ for both Pr = 7.0 and 11.4. However, the decrease is more rapid in water at room temperature. Greater viscous dissipative heat also causes a fall in the value of $\{-f''(0)\}$ when $Gr/Re^2 > 0$ whereas opposite is the case when $Gr/Re^2 < 0$. Again in the presence of opposing flows, an increase in Gr/Re^2 (<0), leads to a rise in the values of $\{-f''(0)\}$. Again, when $Gr/Re^2 > 0$, the rate of heat transfer values of $\{-\theta'(0)\}$ increase with increasing Gr/Re^2 (>0) for both Pr = 7.0 & Pr = 11.4. Opposite is the case when $Gr/Re^2 < 0$ is increased. Also when Gr/Re^2 (<0), greater viscous dissipative heat causes a fall in the values of $\{-\theta'(0)\}$.

Table 1. Values of $\{-f''(0)\}$ and $\{-\theta'(0)\}$

Pr	$\frac{Gr}{Re^2} \backslash E$	0.2	0.5	0.2	
7.0	0.2	0.3543		1.1822	
	0.4	0.2662		1.2095	
	0.6	0.1794		1.2327	
11.4	0.2	0.4000	0.3878	1.5118	1.0842
	0.4	0.3563	0.3324	1.5246	1.12313
	0.6	0.3128		1.5365	
	-0.2	0.4882	0.5007	1.4835	0.9972
	-0.4	0.5327	0.5583	1.4679	0.9468
	-0.6	0.5776		1.4513	

CONCLUSIONS

a) Velocity increases with increasing Gr/Re^2 (> 0) and decreases with increasing Gr/Re^2 (<0)
b) Due to more viscous dissipative heat, velocity increases.
c) Temperature decreases with increasing Gr/Re^2 (>0) and when Gr/Re^2 (<0), it increases with increasing Gr/Re^2 (<0).
d) Due to greater viscous dissipative heat there is a rise in the temperature of water at 4^0C when Gr/Re^2 (>0), but temperature is not significantly affected when Gr/Re^2 (<0).
e) The values of $\{-f''(0)\}$ decrease with increasing Gr/Re^2 (>0) and with greater viscous dissipative heat, opposite is the case when Gr/Re^2 (<0).
f) The values of $\{-\theta'(0)\}$ increases with increasing Gr/Re^2 (>0) and opposite is the case when Gr/Re^2 (<0).
g) When Gr/Re^2 (<0), greater viscous dissipative heat causes a fall in the value of $\{-\theta'(0)\}$.

NOMENCLATURE

Nomenclature

E	Eckert number
f	dimensionless stream function
g	acceleration due to gravity
Gr	Grashof number
K	thermal conductivity
Nu	Nusselt number
Pr	Prandtl number
T	Temperature
u, v	velocity components along radial and axial directions
U_0	reference velocity
α	thermal diffusivity of the medium
β	volumetric coefficient of thermal expansion
θ	dimensionless temperature
μ	coefficient of viscosity
ν	kinematic viscosity
ρ	density
ψ	dimensionless stream function
τ	skin friction

Subscripts

w, ∞ conditions at the wall and in the ambient fluid

Superscript

$/$ prime denotes derivatives with respect to η

REFERENCES

Acrivos A. (1958): Combined laminar free and forced convection heat transfer in external flows. *Trans. Amer. Inst. Chem. Eng.*, Vol. 4, pp. 285-289.

Blasius H. (1908): Greenzschichten in Flussigkeit-enmit Klenier Reibung. *Z.Math. Phys.* Vol. 56, pp. 1-15.

Goren S.L. (1966): On free convection in water at 4^0C. *Chem. Eng. Sci.*, Vol. 21, pp. 515-520.

Howarth L. (1936): Velocity and temperature distribution for a flow along a flat plate. *Proc. Roy. Soc.* (London), Vol. A 154, pp. 364-377.

Howarth L. (1938):On a solution of the laminar boundary layer equations. *Proc. Roy. Soc.* (London), Vol. A 164, pp. 547-579.

Kliegel J.R. (1959): *Combined laminar free and forced convective heat transfer from a vertical flat plate*. Ph.D. Thesis, Univ. of California, Los Angeles, CA.

Sakiadis B.C. (1961): Boundary layer behaviour on continuous solid surfaces II, The boundary layer on a continuous flat surface. *Amer. Inst. Chem. Eng.* Vol. 7, pp. 221-225.

Soundalgekar V.M. and Ramana Murty T.V. and Vighnesam N.V. (1984): Combined forced and free convective flow of water at 4^0C past a semi-infinite vertical plate. *Intl. J. Heat and Fluid Flow*, Vol. 5, pp. 54-56.

Soundalgekar V.M. and Takhar H.S. (1999): On flow past a continuously moving semi-infinite vertical plate in the upward direction. *Applied Mech. And Engng.*, Vol. 4, pp. 553-559.

Sparrow E.M. and Gregg J.L. (1959b): Buoyancy effects in forced convection flow and heat transfer. *J. Appl. Mech.* (Tr. ASME), Vol. 81E, pp. 133-134.

Sparrow E.M., Eichhorn R. and Gregg J.L. (1959a): Combined forced and free convection in a boundary layer flow. *Phys. Fluids*, Vol. 2, pp. 319-328.

Szewczyk A.A. (1964): Combined forced and free convection laminar flow. *J. Heat Transfer*, (Tr. ASME), Vol. 86C, pp. 501-507.

Tsou F.K., Sparrow E.M. and Goldstein R.J. (1967): Flow and heat transfer in the boundary layer on a continuous moving surface. *Intl. J. Heat Mass Transfer*, Vol. 10, pp. 219-229.

In: Perspectives in Mathematical Physics
Editor: Gregory D. Monteggo

ISBN: 978-1-61122-984-4

On the MHD Flow of a Polar Fluid Past an Infinite Plate with Constant Suction

N. Chaturvedi
Department of Mathematics, University of Botswana
Private Bag 0022, Gaborone Botswana

Abstract

The two dimensional polar fluid flow past an infinite porous plate with constant suction has been considered. Expression for the velocity and the skin friction have been obtained in dimensionless forms exponentially increasing and exponentially decreasing cases. It has been observed that the velocity and the skin friction decrease when there is an increase in couple stress parameters.

Keywords: Polar fluid, MHD flow, skin-friction, couple stress, moment of inertia, suction

Introduction

The effect of the magnetic field on the flow of an incompressible and conducting fluid past an infinite plate has been studied by the author (1996). An analysis of the MHD flow of a conducting fluid past a plate in presence of radiation has been studied by Rapits and Massalas (1997). Kinyanjui et al (1999) have studied the MHD stokes problem for a free convection fluid in presence of a strong magnetic field.

Ogulu et al (2002) studied the ubsteady free convection and mass transfer of a fluid past an infinite plate in the presence of thermal diffusion. Author (2003) has also considered the case of MHD Stokes problem for a vertical plate in a heat generating fluid. Several other researchers have done problems regarding the flow past an infinite porous plate where the flow field is governed by the Navier-stokes equations. In the present article, the modified Navier-Stokes equations after taking into account the angular velocity of the fluid particles, have been used. The object of the present analysis here is to study the effects of the properties of the polar fluid on the flow field and the skin-friction.

EQUATIONS OF MOTION

We consider a two-dimensional flow of polar fluid along an infinite porous plate. The flow is independent of the distance parallel to the wall and the suction velocity v'_0 normal to the wall, is directed towards it and is constant. The x' –axis is taken along the wall, y' –axis normal to the wall. Dashes denote the dimensional quantities. Under these assumptions, the physical variables are functions of y' and t' and the unsteady two dimensional equations of motions are:

Momentums

$$\frac{\partial u'}{\partial t'} + v' = -\frac{1}{\rho}\frac{\partial p'}{\partial x'} + (\nu + \nu_r)\frac{\partial^2 u'}{\partial y'^2} + 2\nu_r \frac{\partial \omega'}{\partial y'} \tag{1}$$

$$\frac{\partial w'}{\partial t'} + v'\frac{\partial w'}{\partial y'} = \frac{\gamma}{I}\frac{\partial^2 \omega'}{\partial y'^2} \tag{2}$$

Continuity

$$\frac{\partial v'}{\partial y'} = 0 \tag{3}$$

where u' and v' are the velocity components along x' and y' directions respectively, ν the kinematic viscosity, ν_r the kinematic rotational viscosity, ω' the mean angular velocity of the particles, I a scalar constant of dimension equal to that of moment of inertia of unit mass, and

$$\gamma = \frac{c_a + c_d}{I}$$

where c_a and c_d are the coefficients of the couple stress viscosities.

The boundary conditions are given as;

$$u' = 0, \frac{\partial w'}{\partial y'} = -\frac{\partial^2 u'}{\partial y'^2} \text{ at } y' = 0 \tag{4}$$

$$u' \to U'(t'), w' \to 0 \text{ as } y' \to \infty$$

where $U'(t')$ is the velocity outside the boundary layer. These boundary conditions are derived on the assumption that the couple stresses are dominant during the rotation of the particles.

Now we restrict consideration to the case of v' equal to the negative constant ($-v_0$) where v_o is a non zero negative mean suction velocity directed towards the plate. For free stream we have from (1),

$$-\frac{1}{\rho'}\frac{\partial v'}{\partial y'}=\frac{dU'}{dt'} \tag{5}$$

In view of the above, equations (1) and (2) become,

$$\frac{\partial u'}{\partial t'}-v_0'\frac{\partial u'}{\partial y'}=\frac{dU'}{dt'}+(\upsilon+\upsilon_r)\frac{\partial^2 u'}{\partial y'^2}+2\upsilon_r\frac{\partial \omega'}{\partial y'} \tag{6}$$

$$\frac{\partial \omega'}{\partial t'}-v_0\frac{\partial \omega'}{\partial y'}=\frac{\gamma}{I}\frac{\partial^2 \omega'}{\partial y'^2} \tag{7}$$

subject to the boundary conditions (4).
Now, the non dimensional quantities are introduced defined by,

$$y=\frac{y'v_0'}{\upsilon},\ t=\frac{v_0'^2t'}{4\upsilon},\ n=\frac{4\upsilon n'}{v_0'^2},\ u=\frac{u'}{U_0'},\ U=\frac{U'}{U_0'}$$
$$\omega=\frac{\omega\upsilon}{U_0'v_0'},\ \beta=\frac{I\upsilon}{\gamma},\ \alpha=\frac{\upsilon_r}{\upsilon} \tag{8}$$

where U_0' is the reference velocity. With the help of (8), equations (6) and (7) take the non-dimensional form,

$$\frac{1}{4}\frac{\partial u}{\partial t}-\frac{\partial u}{\partial y}=\frac{1}{4}\frac{dU}{dt}+(1+\alpha)\frac{\partial^2 u}{\partial y^2}+2\alpha\frac{\partial \omega}{\partial y} \tag{9}$$

$$\frac{1}{4}\frac{\partial \omega}{\partial t}-\frac{\partial \omega}{\partial y}=\frac{1}{\beta}\frac{\partial^2 \omega}{\partial y^2} \tag{10}$$

The corresponding boundary conditions are;

$$u = 0, \frac{\partial \omega}{\partial y} = -\frac{\partial^2 u}{\partial y^2}, \quad at \quad y = 0$$

$$u \to U(t), \ \omega \to 0, \quad as \quad y \to \infty \tag{11}$$

Here α and β are new parameters which help us to study the effects of the couple stresses on the flow field.

Now we will solve equations (9) and (10) with the boundary condition (11) in two cases; (a) when the free stream velocity follows the exponentially increasing law and (b) when it follows the exponentially decreasing law.

Exponentially Increasing Case

In this section, we consider the case in which the free stream velocity follows the exponentially increasing small perturbation law.

Let us assume therefore that in the free stream,

$$U = 1 + \varepsilon e^{nt} \tag{12}$$

and in the neighbourhood of the plate,

$$u = f_1(y) + \varepsilon e^{nt} f_2(y) \tag{13}$$

and

$$\omega = F_1(y) + \varepsilon e^{nt} F_2(y) \tag{14}$$

The above assumptions are restricted to (i) time t is finite (ii) nt is small such that $0 < nt < 1$ (iii) ε is a small positive parameter so that εe^{nt} is convergent.

Substituting (12),(13) and (14) in (9) and (10) , equating the harmonic and non harmonic terms and neglecting the coefficients of ε^2, we get

$$(1+\alpha) f_1'' + f_1' = -2\alpha F' \tag{15}$$

$$(1+\alpha) f_2'' + f_2' - \frac{n}{4} f_1 = -\frac{n}{4} - 2\alpha F_2' \tag{16}$$

$$F_1'' + \beta F_1' = 0 \tag{17}$$

$$F_2'' + \beta F_2' - \frac{n\beta}{4} F_2 = 0 \tag{18}$$

Here, primes denote the derivatives with respect to y. These are coupled differential equations which are difficult to solve. Hence we again assume,

$$f_1 = f_{11} + \alpha f_{12}, \quad f_2 = f_{21} + \alpha f_{22}, \qquad (19)$$
$$F_1 = F_{11} + \alpha F_{12}, \quad F_2 = F_{21} + \alpha F_{22}$$

These assumptions are possible only when $\alpha < 1$ i.e when $\upsilon_r < \upsilon$. Substituting (19) in (15) to (18), equating the coefficients of different powers of α, neglecting those of α^2, we get

$$f''_{11} + f'_{11} = 0 \qquad (20)$$

$$f''_{12} + f'_{12} = -f''_{11} - 2F'_{11} \qquad (21)$$

$$f''_{21} + f'_{21} - \frac{n}{4} f_{21} = -\frac{n}{4} \qquad (22)$$

$$f''_{22} + f'_{22} - \frac{n}{4} f_{22} = -2F'_{21} - f''_{21} \qquad (23)$$

$$F''_{11} + \beta F'_{11} = 0 \qquad (24)$$

$$F''_{12} + \beta F' = 0 \qquad (25)$$

$$F''_{21} + \beta F'_{21} - \frac{n\beta}{4} F_{21} = 0 \qquad (26)$$

$$F''_{22} + \beta F'_{22} - \frac{n\beta}{4} F_{22} = 0 \qquad (27)$$

The corresponding boundary conditions are,

$$f_{11}=0,\ f_{12}=0,\ f_{21}=0,\ f_{22}=0,\ F'_{11}=-f''_{11},\ F'_{12}=-f''_{12},\ F'_{21}=-f''_{21},\ F'_{22}=-f''_{22} \quad at\ y=0$$

and

$$f_{11} \to 1,\ f_{12} \to 0,\ f_{21} \to 1,\ f_{22} \to 0,\ F_{11}, F_{12}, F_{21}, F_{22} \to 0 \quad as\ y \to \infty \qquad (28)$$

$$f_1 = 1 - e^{-y} + \frac{2\alpha}{\beta^2 - \beta}\left(e^{-y} - e^{-\beta y}\right) - \alpha\, y e^{-y} \qquad (29)$$

$$f_2 = 1 - e^{-y} + \frac{2 l_1^2 \alpha}{s_1^2 - s_1 - \frac{n}{4}}\left(e^{-l_1 y} - e^{-s_1 y}\right) - \frac{l_1^2 \alpha}{2 l_1 - 1}\, y e^{-y} \qquad (30)$$

$$F_1 = \left[\frac{2\alpha(1-\beta^2)}{\beta(\beta^2-\beta)} + \frac{(2\alpha-1)}{\beta}\right] e^{-\beta y} \qquad (31)$$

$$F_2 = \left[\frac{2 l_1^3 \alpha}{s_1(2l_1 - 1)} - \frac{2 l_1^3 \alpha (s_1^2 - l_1^2)}{s_1\left(s_1^2 - s_1 - \frac{n}{4}\right)} - \frac{l_{1s}^2}{s_1}\right] e^{-s_1 y} \qquad (32)$$

where

$$l_1 = \frac{1 + \sqrt{1+n}}{2}, \quad s_1 = \frac{\beta + \sqrt{\beta^2 + n\beta}}{2}$$

Equations (21) to (28) are now solved subject to the above boundary condition and substituted in (20) to give,

Hence the velocity in the boundary layer, is given by

$$u(y,t) = 1 - e^{-y} + \frac{2\alpha}{\beta^2 - \beta}(e^{-y} - e^{-\beta y}) - \alpha y e^{-y} + \in e^{nt}\left[1 - e^{-l_1 y} + \frac{2 l_1^1 \alpha}{s_1^2 - s_1 - \frac{n}{4}}(e^{-l_1 y} - e^{s_1 y}) - \frac{l_1^2 \alpha}{2l_1 - 1} y e^{-y}\right] \quad (33)$$

$$\omega(y,t) = \left[\frac{\alpha(1-\beta^2)}{\beta(\beta^2 - \beta)} + \frac{(2\alpha - 1)}{\beta}\right] e^{-\beta y} + \in e^{nt}\left[\frac{2 l_1^3 \alpha}{s_1(2l_1 - 1)} - \frac{2 l_1^2 \alpha (s_1^2 - l_1^2)}{s_1(s_1^2 - s_1 - \frac{n}{4})} - \frac{l_1^2}{s_1}\right] e^{-s_1 y} \quad (34)$$

The non dimensional skin-friction at the plate is given by

$$\tau_0 = \left(\frac{\partial u}{\partial y}\right)_{y=o} = (1-\alpha) + \frac{2\alpha}{\beta} + \varepsilon e^{nt}\left[l + \frac{2 l_1^2 \alpha (s_1 - 1)}{s_1^2 - s_1 - \frac{n}{4}} - \frac{l_1^2 \alpha}{2l_1 - 1}\right] \quad (35)$$

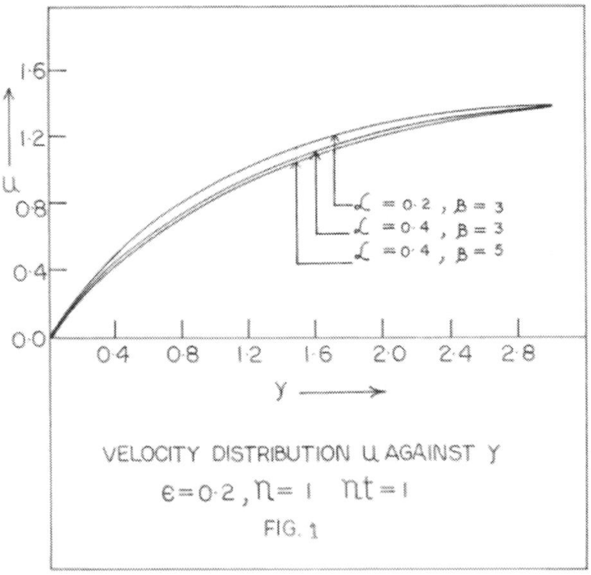

Figure 1.

Figure 1 shows the variation of the velocity field u with regard to y for $\in = 0.2$, n =1, nt = 1 and for different values of the parameters as shown in the figure. It is clear from the figure that the velocity u decreases when for the same value of α, there is an increase in β. Also, u decreases when for the same values of β , there is an increase in α. It amounts to say that the velocity decreases either by increasing α or β or both. In Figure, the skin –friction τ_0 is shown against α. From this graph, we conclude that the skin-friction goes on decreasing by increasing the parameters α and β. Thus, we can say that the velocity and skin-friction decrease when the kinetic rotational viscosity ν_r increases.

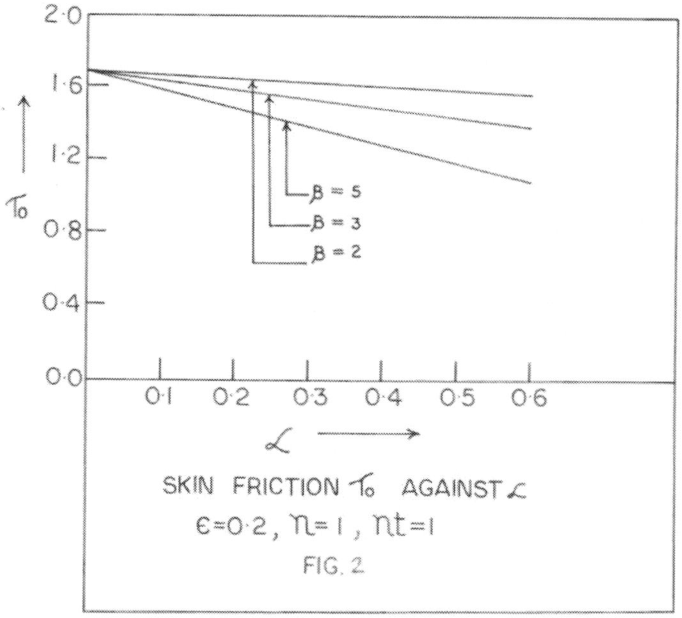

Figure 2.

Exponentially Decreasing Case

We now consider the case when the external flow velocity follows the exponentially decreasing small perturbation law.

Let the free stream velocity U and the velocity in the boundary layer be given by

$$U = 1 + \varepsilon\, e^{-nt} \tag{36}$$

and

$$u = f_1(y) + \varepsilon\, e^{-nt} f_2(y) \tag{37}$$

$$\omega = F_1(y) + \varepsilon e e^{-nt} F_2(y) \tag{38}$$

In this case as well the boundary conditions are the same. Now, following the same method as described in the first part, we obtain

$$f_1 = 1 - e^{-y} + \frac{2\alpha}{\beta^2 - \beta}(e^{-y} - e^{-\beta y}) - \alpha y e^{-y} \tag{39}$$

$$f_2 = 1 - e^{-l_2 y} + \frac{2\alpha l_2^2}{s_2^2 - s_2 + \frac{n}{4}}(e^{-l_2 y} - e^{-s_2 y}) - \frac{\alpha l_2^2}{2l_2 - 1} y e^{-l_2 y} \tag{40}$$

$$F_1 = \left[\frac{2\alpha(1-\beta^2)}{\beta(\beta^2 - \beta)} + \frac{2\alpha - 1}{\beta}\right] e^{-\beta y} \tag{41}$$

$$F_2 = \left[\frac{2\alpha l_2^5}{s_2(2l_2 - 1)} - \frac{2\alpha l_2^3(s_2^2 - l_2^2)}{s_2(s_2^2 - s_2 + \frac{n}{4})} - \frac{l_2^2}{s_2}\right] e^{-s_2 y} \tag{42}$$

where

$$l_2 = \frac{1 + \sqrt{1-n}}{2}, \qquad s_2 = \frac{\beta + \sqrt{\beta^2 - n\beta}}{2}$$

Hence the velocity field in this case is given

$$u(y,t) = 1 - e^{-y} + \frac{2\alpha}{\beta^2 - \beta}(e^{-y} - e^{-\beta y}) - \alpha y e^{-y} + e\, e^{-nt}\left[1 - e^{-l_2 y} + \frac{2\alpha l_2^2}{s_2^2 - s_2 + \frac{n}{4}}(e^{-l_2 y} - e^{-s_2 y}) - \frac{\alpha l_2^2}{2l_2 - 1} y e^{-l_2 y}\right] \tag{43}$$

and

$$\omega(y,t) = \left[\frac{2\alpha(1-\beta^2)}{\beta(\beta^2 - \beta)} + \frac{2\alpha - 1}{\beta}\right] e^{-\beta y} + \varepsilon e^{-nt}\left[\frac{2\alpha l_2^3}{s_2(2l - 1)} - \frac{2\alpha l_2^2(s_2^2 - l_2^2)}{s_2(s_2^2 - s_2 + \frac{n}{4})} - \frac{l_2^2}{s_2}\right] e^{-s_2 y} \tag{44}$$

The skn-friction τ_0 at the plate is,

$$\tau_0 = \left(\frac{\partial u}{\partial y}\right)_{y=0} = (1-\alpha) + \frac{2\alpha}{\beta} + \varepsilon e^{-nt}\left[l_2 + \frac{2\alpha l_2^2}{s_2^2 - s_2 + \frac{n}{4}}(s_2 - 1) - \frac{\alpha l_2^2}{2l_2 - 1}\right] \tag{45}$$

Figure 3 and Figure 4 exhibit the effects of the parameters α and β on the velocity field u and the skin- friction τ_0. In figure (3), the velocity increases as y increases but it decreases when there is an increase in α or β or both. The skin-friction as shown in Figure 4 decreases when α and β increase. Thus the variation in the velocity and skin-friction in this case is similar to that of the previous case.

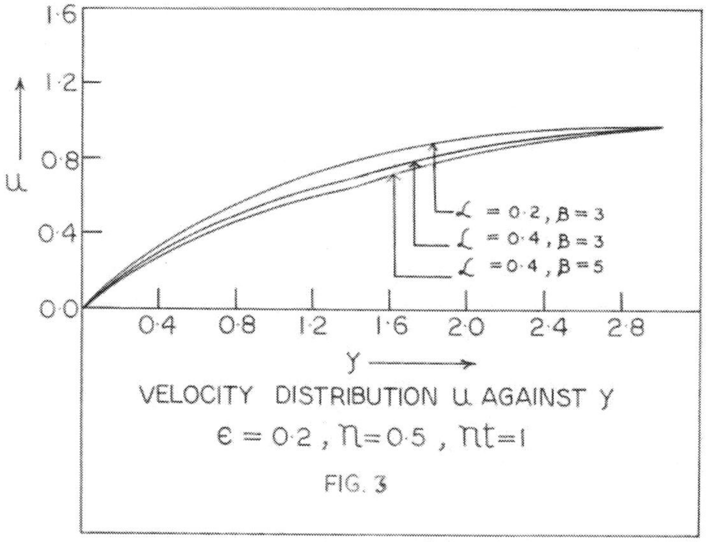

Figure 3.

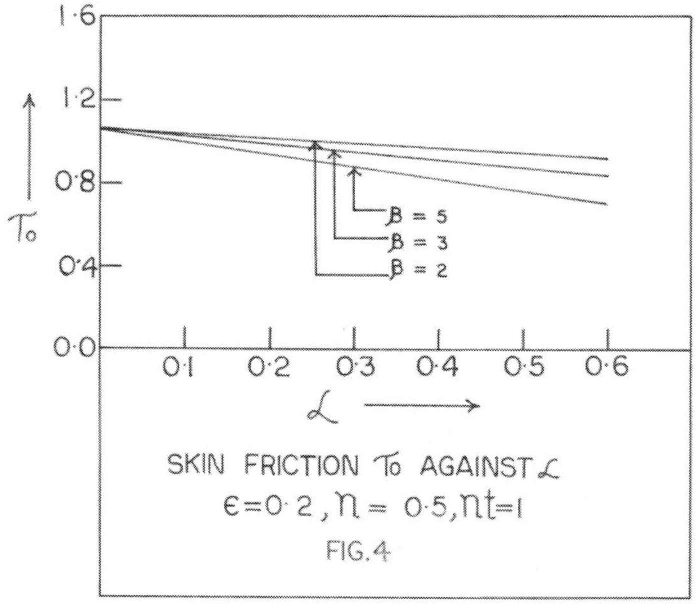

Figure 4.

Conclusions

i. The velocity and the skin-friction decrease when there is an increase either in α or β or both.

ii. For the same value of α, the skin-friction in the exponentially increasing case is greater than that of exponentially decreasing case.

References

Chaturvedi, N.; *Bots.Jr. of Tech.*, 12, 2, 21-31, (2003)
Chaturvedi, N.; *Energy Convers. Mgmt*, 37, 5, 623-627, (1996)
Chaturvedi, N.; *Energy. Convers.Mgmt.* 30, 49, (1990)
Kinyanjui,M et al.; *Jr.of MHD and Plasma Res.* 8, 4, 301-319, (1999)
Ogulu,A et al.; *Bots.Jr.of Tech.*, 11,1, 52-56, (2002)
Rapits,A. and Massila,C.V.; *Jr.of MHD and Plasma Res.*7, 2/3,121-123, (1997)

In: Perspectives in Mathematical Physics
Editor: Gregory D. Monteggo
ISBN 978-1-61122-984-4

Chapter 5

QUANTUM-MAGNETIC COLLAPSE OF A NEUTRON GAS: NO MAGNETAR FORMATION

***Herman J. Mosquera Cuesta*[a,d], *Aurora Pérez Martínez*[b,d] and *Hugo Pérez Rojas*[b,c,d]**

[a]Instituto de Cosmologia, Relatividade e Astrofísica (ICRA-BR), Centro Brasileiro de Pesquisas Físicas
Rua Dr. Xavier Sigaud 150, Cep 22290-180, Urca, Rio de Janeiro, Brazil
[b] Grupo de Física Teórica, ICIMAF, Calle E No. 309, 10400 La Habana, Cuba
[c] High Energy Physics Division, Department of Physics, University of Helsinki and Helsinki Institute of Physics, P. O Box 64, FIN-00014 Helsinki, Finland
[d] Abdus Salam International Centre for Theoretical Physics, P.O. Box 586, Strada Costiera 11, Miramare 34100, Trieste, Italy

Abstract

The origin and early evolution of super strongly magnetized neutron star pulsars with magnetic fields on the order of 10^{14-15} Gauss, the so-called magnetars, has been under focus in the late years. The suggestion that these objects could appear during the initial evolution of a neutron star, as a consequence of the amplification of the star's remnant magnetic field via the $\alpha - \Omega$ dynamo mechanism, has stimulated an intense research activity. Observations of the Soft Gamma-ray Repeaters (SGRs), a subclass of astrophysical sources emitting powerful 'soft' gamma-ray bursts; with mean energies of 10^{40-44} erg s and $\sim 3 - 10$ years time recurrence, have led to the claim that the magnetars may have been discovered. Meanwhile, to put those theoretically classical objects on realistic astrophysical grounds is needed to incorporate to their physical description also the quantum effects. This could be done in the first approximation by superimposing them to the star gravity. In this paper we study a model of a neutron star envisioned as a relativistic degenerate neutron gas in equilibrium with a background of electrons and protons acted upon by a magnetic field. It is shown that this system exerts its pressure anisotropically, having a smaller value perpendicular to, instead of along, the magnetic field. For critical fields, as those said typical of magnetars, the magnetic pressure may produce the vanishing of the transversal (to the magnetic field) pressure of the neutron gas, because of its ferromagnetic behavior. Taking it as a model of a neutron star, the outcome could be a transverse collapse of the star. This quantum effect imposes a limit to the fields to be observable in stable neutron star pulsars as a function of their density. The final structure left over after the implosion might be a mixed phase of nucleons and meson condensate, a strange star, or a black string, but no a magnetar at all.

1 Introduction

Gravitational collapse occurs in a body of mass M and radius R when its rest energy is of the same order of its gravitational energy, i. e., $Mc^2 \sim GM^2/R$. We would like to argue that for a macroscopic magnetized body, e.g., composed by neutrons in an external field $\vec{B}$ ($|\vec{B}| \equiv B$), new physics arises and a sort of collapse occurs when its internal energy density U is of the same order than its magnetic energy density $\vec{\mathcal{M}} \cdot \vec{\mathcal{B}}$, where $\vec{\mathcal{M}}$ is the magnetization. This problem is interesting in the context of both cosmology and astrophysics, as for instance in the study of objects such as neutron stars. A gas of neutral particles having an anomalous magnetic moment (as a model for neutron stars (NSs); here we assume, as usually, a background of electrons and protons in β equilibrium, which is demanded by Pauli's Principle to guarantee neutron stability), when placed in extremely strong magnetic fields has a nonlinear (ferromagnetic) response to the external field and is also unstable due to the vanishing of the transverse pressure for surface fields strong enough ($B_{surf} \geq 10^{16}$ G). In this phenomenon quantum effects play an essential role due to the coupling of the particles' spins to the microscopic field $\vec{B}$ seen by the particles (spin-polarization effect). For fields of this order of magnitude there are values of the density for which the magnetic energy of the system becomes of the same order of magnitude than its total energy. At these physical conditions any structure of super dense matter composed of neutral particles having a magnetic moment may undergo a transverse collapse when its pressure perpendicular to $\vec{B}$ vanishes. This implosion is driven by the same mechanism described in [1] for charged particles.

We present in this new paper, which is a more elaborated version of [2], the main ideas concerning the role of ultra strong magnetic fields in a gas of neutral particles: The standard model of NSs, as envisioned by Duncan and Thompson (D&T) in their model of magnetars[3], based on the D&T model. The fundamental result that we obtain shows that a NS, i. e., a neutron gas permeated by a super strong magnetic field is unstable and must collapse. This inedit result seems to ban the possibility of formation magnetars. We stress in this respect the fact that similar results were obtained by two different groups: Khalilov [4] and Ghosh, Mandal and Chakrabarty [5]. Although both teams of researchers arrived to the same conclusion like ours, quite different approaches were pursued.

The paper is organized as follows: Section II reviews the concept of anisotropic pressure in self-gravitating systems like NSs. In Section III are given the tools to construct the energy-momentum tensor of a neutron gas. Section IV discusses the main differences between the classical and quantum collapse of a gas configuration in an approximation-independent way, based on the sign of the electromagnetic response of the medium to the external applied field. In Section V the dynamics of the neutron gas composing a NS is studied in the presence of a magnetic field, followed in Section VI by the derivation of its thermodynamical potential and magnetization. Section VII discusses the conditions for the collapse to take place, and an application of this physics to the stability analysis of the proposed magnetars is given. The main result of this paper shows that such objects should not form if they are envisioned as a standard neutron gas in Fermi beta equilibrium as is claimed in the original idea introducing the concept of magnetar. Some closing remarks and further potential applications of this theory are part of the final section.

2 Anisotropic Pressures in Self-gravitating Systems

The issue of local anisotropy in pressures was extensively reviewed by Herrera and Santos [6] in a general relativistic approach. These authors present several physical mechanisms for its origin in both extremely low and very high density systems, which may include astrophysical compact objects. In the case of highly dense systems, it was pointed out that "exotic" phase transitions could occur during gravitational collapse, which is exactly the problem we are concerned with in this paper. However, the more fundamental problem regarding the appearance of anisotropic pressures in strongly magnetized compact stars was left open. More recently, Mak and Harko [7] present a class of exact solutions of Einstein's equations corresponding to anisotropic stellar configurations which can describe realistic neutron stars.

We want to provide a more detailed description of the arising of anisotropic pressures in a relativistic system like a NS, an essential point in understanding the physics behind the problem of stability of ultra magnetized compact stars. We shall give firstly general arguments to support this our view, and then we concentrate in the specific calculations, in the one-loop approximation, of the thermodynamic potential of the neutron star configuration and its magnetization, the properties upon which the most crucial conclusions can be drawn.

To fix ideas, we shall work in the grand canonical ensemble, and we are considering as subsystem, some region inside the star. Such subsystem is under the influence of the magnetic field created by the rest of the system, which we name $\vec{H}$ ($|\vec{H}| \equiv H$). The response of the subsystem is to polarize itself creating a magnetization $\vec{\mathcal{M}}$ in the medium (the neutron gas) satisfying the relation: $\vec{H} = \vec{B} - 4\pi\vec{\mathcal{M}}$. Obviously, inside the subsystem the microscopic field is $\vec{B} = \vec{H} + 4\pi\vec{\mathcal{M}}$, since $\vec{B}$ (named also magnetic induction) and the electric field $\vec{E}$ are the *true fields* acting on the electric charges and magnetic dipoles of elementary particles [8] [9]. (Note, however, that Landau in Ref.[10] uses the notation $\vec{H}$ to name the magnetic field in vacuum). The field $\vec{B}$, as well as $\vec{E}$, satisfy the Maxwell equations for particles in vacuum. In what follows, when we are to discuss the dynamics of the particles in the neutron gas we will sometimes refer to $\vec{B}$ as the *external magnetic field*, as is usually named in quantum field theory and astrophysics. For an external distant observer, $\vec{B} = \vec{H}$ (in Gaussian units) since the magnetization is assumed to exist only inside the star. We emphasize that actually $\vec{B}$ and $\vec{H}$ are external fields within different contexts: $\vec{H}$ is external to the *subsystem* object of study in the grand canonical ensemble, whereas $\vec{B}$ is external to any particle chosen in the subsystem (it feels, in addition to $\vec{H}$, the contribution from the magnetization field $4\pi\vec{\mathcal{M}}$ due to the other particles of the subsystem).

In the case of a gas of electrically charged particles in an external constant magnetic field $\vec{B}$, in classical electrodynamics, it is the Lorentz force $\vec{F} = e\vec{v} \times \vec{B}/c$ the source of an asymmetry in the pressure components parallel and perpendicular to $\vec{B}$. By writing $e\vec{v} = \vec{j}\Delta V$, where $\Delta V = dx_1 dx_2 dx_3$, calling $f_i = F_i/\Delta V$ as the i-th component of the force density, and substituting $\vec{j} = c\nabla \times \vec{\mathcal{M}}$, one has

$$f_i = -(\partial_i \mathcal{M}_s)B_s + (\partial_s \mathcal{M}_i)B_s \ . \tag{1}$$

By multiplying by $\Delta V = dx_1 dx_2 dx_3$ and assuming $B_s = B\delta_{s3}$ and $\partial \mathcal{M}_i/\partial x_3 = 0$ (actually it is also $\mathcal{M}_i = \mathcal{M}\delta_{i3}$), only the first term in (1) remains nonzero and one recovers an expression for the Lorentz force, which is obviously perpendicular to the field $\vec{B}$. For the corresponding pressure it yields $P_{L\perp} = -\vec{\mathcal{M}} \cdot \vec{B}$. This is a classical effect and obviously $P_{L\perp}$ must be added to the usual kinetic isotropic pressure P_0, so that the total transverse pressure becomes $P_\perp = P_0 + P_{L\perp}$. As in classical electrodynamics, by Lenz law, $\vec{\mathcal{M}}$ is opposite to $\vec{B}$ (spin effects are neglected), then $\vec{\mathcal{M}} \cdot \vec{B} < 0$, and $P_{L\perp} > 0$. The opposite case occurs when $\vec{\mathcal{M}}$ is parallel to $\vec{B}$, which occurs in the quantum case, i.e., when spin effects are taken into account. We alert, however, that in the definitions and derivations that follows use will be done of the magnitudes $|\vec{B}|$ and $|\vec{\mathcal{M}}| \equiv \mathcal{M}$ of both vectors $\vec{B}$ and $\vec{\mathcal{M}}$ instead of the vectors themselves.

3 The Energy-Momentum Tensor of a Neutron Gas

Based on more fundamental grounds, one may write the general structure of the energy-momentum tensor of a neutron gas in an external field $\vec{B}$ in the same way as one does to construct general tensors, as the polarization operator tensors [11], for instance. In an external field $F_{\mu\nu}$, in addition to the basic 4-velocity vectors of the medium, u_μ, and particle momentum k_μ, we have two extra vectors $F_{\mu\nu}k_\nu$, $F^2_{\mu\nu}k_\nu$, to form a basis of independent vectors (in what follows we shall use the notation $F^2_{\mu\nu} = F_{\mu\lambda}F_{\lambda\nu}$) . From them we may build a set of basic tensors which, together with the tensors $\delta_{\mu\nu}$, $F_{\mu\nu}$, $F^2_{\mu\nu}$, serves as a basis in terms of which we can expand any tensor structure related to the particle dynamics, in particular, the energy-momentum tensor. However, to get rid of tensor structures containing off-diagonal terms, which would correspond to unwanted shearing stresses in the fluid rest frame $u = (0, 0, 0, u_4)$, we exclude some of them, i. e., $k_\mu k_\nu$, $k_\mu F_{\nu\lambda}k_\lambda$, $F_{\mu\nu}\, u_\mu k_\nu$, or any of its combinations. By following the arguments used in [11], we conclude that we are left in the present case with three basic tensors: $\delta_{\mu\nu}$, $F^2_{\mu\nu}$, $u_\mu u_\nu$ to describe the dynamics of such a neutron gas. Thus, the structure of such an energy-momentum tensor is then expected to be of the form

$$\mathcal{T}_{\mu\nu} = a\delta_{\mu\nu} + bF^2_{\mu\nu} + cu_\mu u_\nu \tag{2}$$

where $\mu, \nu = 0, 1, 2, 3$; and $a = p$ is the isotropic pressure term, $b = \mathcal{M}/B$ and $c = U + p$. In the present case the second of these tensors can be written in a simpler form as $F^2_{\mu\nu} = -B^2\delta^\perp_{\mu\nu}$. The tensor (2) has then the spatial eigenvalues $p - B\mathcal{M}, p - B\mathcal{M}, p$, and the time eigenvalue $-c = -U - p$, since $u_\mu u_\mu = -1$. These eigenvalues exhibit the anisotropy in pressures perpendicular and parallel to B.

In dealing with a quantum gas in an external field we shall assume that the sources of the field $\vec{H}$ are either classical (currents) or due to quantum effects. Although it is out of the scope of this paper to discuss the mechanism for producing such field, we suggest, however, some viable sources able to induce a self-consistent field, as for instance a condensate of the vector meson ρ, the neutron spin-spin ferromagnetic [12] coupling (see below), or even by diquark [13] condensation.

As in the case of the electron in an atom the basic dynamics in our present case is described by the Dirac equation in an external field, in place of the Lorentz force. According to the Ehrenfest theorem, classical dynamics is contained on the adequate average of quantum dynamics. But quantum dynamics leads also to several new phenomena not having classical partner. After solving the Dirac equation one gets the energy eigenvalue spectrum [14]. This energy depends on the microscopic magnetic field B through some interaction term in the initial Lagrangian. These energy eigenvalues, after the quantum statistical average, determine the thermodynamic properties, such as the neutron gas pressure. If the coupling constant is turned to zero, the particles would not feel any pressure coming from the external field. Thus, since the classical Maxwell stress tensor of the field H does not express by itself the interaction with the particles, i. e., it expresses the momentum and energy of the external field, we have no need to add it to the expression (4) below. The tensor (4) contains already the basic tensor structures of the problem, including the Maxwell stress tensor of the field B, which depends on the external field H and the magnetization $|\vec{\mathcal{M}}|$.

The total external field B contributes with virtual particles, expressed through the Euler-Heisenberg vacuum terms arising in the regularization of the quantum vacuum terms [15]. These vacuum terms appear in the calculation of the basic statistical quantity, the thermodynamical potential: $\Omega \equiv -\beta^{-1} \ln \mathcal{Z}$, where $\mathcal{Z}$, the grand partition function, is built up on the particle spectrum. Our thermodynamical potential is the sum of two terms, $\Omega = \Omega_{st} + \Omega_V$, the finite statistical term Ω_{st} plus the vacuum field contribution Ω_V, which is divergent. In the process of regularization, it absorbs the classical field energy density term $B^2/8\pi$. We observe here that Landau, in p.69 of Ref.[16], uses the specific term *thermodynamic potential*, denoted by Φ in that reference, to name what in western literature is known as Gibbs free energy, which is denoted by G. Our thermodynamic potential is just what Landau defines as "new thermodynamic potential Ω", but we have taken it per unit volume, which is dependent on T and μ in absence of external field, and in our case is dependent also on B. Observe that $\Omega = F - G$. In the zero field case, it would be $\Omega = -P$, where P is the isotropic pressure. Due to the spatial anisotropy introduced by the magnetic field B, the pressures are not the same in all directions, and only in the direction parallel to B it adquires the value $\Omega(B) = -P_3$ (See below).

The coupling of the spin dipole moment of neutrons in an external magnetic field B produces a loss of rotational symmetry of the particle spectrum (in what follows we will consider B along the x_3 axis). From the spectrum, which is expressed in terms of B, by the standard methods of finite temperature quantum field theory, we obtain the thermodynamical potential (per unit volume) $\Omega = \Omega(B)$, and from it all the thermodynamic properties of the system, in particular its magnetization, as is done by Huang [17] (p. 237), which is the statistical average $\mathcal{M} = -(\partial\Omega/\partial B)$.

Notice that this definition is consistent with our convention by which thermodynamical quantities are defined in terms of the microscopic magnetic field B acting on the particles as an independent variable (see for instance Ref.[9]), instead of using the quantity H. Thus, the thermodynamical variable conjugated to the magnetic field B is the quantity $\mathcal{M}(B)$, the system magnetization as introduced above. We may write then

$$\Omega = -\int \mathcal{M} dB - P_0, \tag{3}$$

where $P_0 = -\Omega(0)$ is the term corresponding to the zero magnetic field pressure. One must emphasize that in the quantum relativistic case Ω depends on B nonlinearly. From the explicit expression for Ω given in another section below, one finds that the dependence of the energy spectrum on the particle momentum is not rotational invariant. This fact determines a reduction of the symmetry of the otherwise isotropic thermodynamic properties of the system such as the pressure, which is expected to be axially-symmetric for the reasons pointed out above.

By using the Green functions method it is found that the energy-momentum tensor of matter in an external constant magnetic field obeys the general structure (2), and we have [1]

$$\begin{aligned} \mathcal{T}_{\mu\nu} &= (T\partial\Omega/\partial T + \sum \mu_r \partial\Omega/\partial\mu_r)\delta_{4\mu}\delta_{4\nu} \\ &+ 4F_{\mu\lambda}F_{\nu\lambda}\partial\Omega/\partial F^2 - \delta_{\mu\nu}\Omega, \end{aligned} \tag{4}$$

where r run over the species involved. Below we will take $r = n, p, e$ to describe the neutron, proton and electron component of the star gas. Expression (4) in the zero field limit reproduces the usual isotropic energy-momentum tensor $\mathcal{T}_{\mu\nu} = P\delta_{\mu\nu} - (P+U)\delta_{4\mu}\delta_{\nu 4}$ of a perfect fluid. From Eq.(4) the spatial components are $\mathcal{T}_{33} = P_3 = -\Omega, \mathcal{T}_{11} = \mathcal{T}_{22} = P_{\perp} = -\Omega - B\mathcal{M}$. The time component $\mathcal{T}_{00} \equiv -U_r = -TS_r - \mu_r N_r - \Omega_r$, in which $U_r \sim \mu_r N_r \sim \Omega_r$ are quantities of the same order of magnitude. We shall assume that inside a stable NS there is locally a balance between the mechanical stresses (and forces) from the pressure exerted by the quantum gas and the gravitational stress, i. e., the gravitational force per unit area, exerted by the star mass, as defined below by equations (6),(7). By abusing a bit of the phrasing we shall refer over the paper to this last one as the "gravitational pressure", in an attempt to turn more meaningful discussions regarding effects of the gas pressure and those stemming from its own gravity. In the case of a neutron star the correct treatment of this problem requires of general relativistic corrections[18]. However, for the general purpose of this paper we consider here the problem in a rough manner, in a similar fashion as it was done in the white dwarf case with zero magnetic field [17], in Newtonian gravity. If dS_i is an element of surface and dx_i an increase in the coordinate $i = 1, 2, 3$, one can express the balance between the work done on the star and the variation of its gravitational energy, $E_g \equiv GM^2R^{-1}$, under a small change of volume ΔV as

$$\int(-\mathcal{T}_{ij}dS_j + \frac{\partial E_g}{\partial x_i})dx_i = 0 , \tag{5}$$

where the quantity in parenthesis must be zero, and describes the balance of forces. We will use now cylindrical coordinates with the z axis parallel to B, and assume that the shape of the deformed star can be (very roughly) approximated by the simplified model of a cylinder of height Z and radius $r_{\perp}$ (we could equally use as a simple model a cylinder of height Z ended by two half-spheres of radius $r_{\perp}$. The equations (6) and (7) below would differ by unimportant numerical factors). We assume also that the change of the cylinder's volume ΔV and surface ΔS are small, so that the total surface of the star $S = 2\pi r_{\perp}(r_{\perp} + Z)$ is approximately constant. Then we have $dS_{\perp} = 2\pi r_{\perp} dz$, $dS_3 = 2\pi r_{\perp} dr_{\perp}$ and integrating we get the equilibrium between the gravitational and gas pressures $P_{\perp} = P_{g\perp}$ and $P_{33} = P_{g33}$, where

$$P_{g\perp} = \frac{1}{2\pi r_\perp Z}\frac{\partial E_g}{\partial r_\perp}. \tag{6}$$

and

$$P_{g33} = \frac{1}{\pi r_\perp^2}\frac{\partial E_g}{\partial z}. \tag{7}$$

Eqs.(6),(7) are to be interpreted respectively as the transverse and longitudinal "gravitational pressures" in the sense defined above. By assuming that $\frac{\partial E_g}{\partial r_\perp}$ and $\frac{\partial E_g}{\partial z}$ are quantities of the same order, to preserve the balance of transverse pressures when $P_\perp$ decreases (due to the increase of the product $B\mathcal{M}$), the quantity $1/2\pi r_\perp Z$ must decrease. This is achieved whenever Z increases faster than $r_\perp$ decreases (all this can be seen by assuming that the total surface $S = const$ and $dr_\perp/d\tau < 0$, where τ is a parameter like time). In that case the outcome is a stretching of the body along the direction of B. Thus, the anisotropy of pressures in our problem suggests that matter in the body is also distributed anisotropically, what leads to prolate isobaric surfaces (This is an effect opposite to the oblateness of the Sun, Earth and other planets due to the effective decrease of the transverse gravitational force by the centrifugal force.).

The condition $P_\perp = 0$ implies also that $U \sim B\mathcal{M}$. As $P_3 > 0$, an instability arises in the system leading to a transverse collapse. Thus, as pointed out in the introduction, new physics arise in connection with the vanishing of the transverse pressure, $P_\perp = 0$. This peculiar behavior will be discussed in the forthcoming sections.

By writing $\mathcal{M} = (B-H)/4\pi$, one may formally write $\Omega = -\frac{1}{8\pi}B^2 + \frac{1}{4\pi}\int HdB - P_0$. We remind that as $\Omega \equiv F - G$, the last expression is consistent with what would be obtained in the classical nonrelativistic case [16], where $F = F_0 + \int HdB/4\pi$ is the Helmholtz free energy and $G = F + \int \mathcal{M}dB + P_0 = G_0 + B^2/8\pi$ as the Gibbs free energy. Because of our definition of $\mathcal{M}$, the last term is given in terms of B and not in terms of H. As we have

$$\mathcal{T}_{33} = -\Omega = \frac{1}{8\pi}B^2 - \frac{1}{4\pi}\int HdB + P_0, \tag{8}$$

and also

$$\mathcal{T}_{11} = \mathcal{T}_{22} = -\Omega - B\mathcal{M} = -\frac{1}{8\pi}B^2 + \frac{1}{4\pi}\int BdH + P_0, \tag{9}$$

it is straightforward to see that the spatial components of the energy-momentum tensor $\mathcal{T}_{ij}$ $(i, j = 1, 2, 3)$ can be rewritten as

$$\mathcal{T}_{ij} = P_0\delta_{ij} - \mathcal{T}^M_{ij}(B, H) + \mathcal{S}(B)_{ij}, \tag{10}$$

where $\mathcal{S}(B)_{ij} = \frac{1}{4\pi}[B_iB_j - \frac{1}{2}(B^2)\delta_{ij}]$ is the Maxwell stress tensor for the microscopic field B, and $\mathcal{T}^M_{ij}(B) = \frac{1}{4\pi}[H_iB_j - (\int BdH)\delta_{ij}]$ is the Minkowski tensor for relativistic nonlinear media (see below). This last term reduces to the usual expression in the nonrelativistic limit if H depends linearly on B [9]. The definition (10) expresses the total pressure as a sum of an isotropic pure mechanical pressure (independent of the electromagnetic field) plus a pressure coming from the Minkowski tensor due to the interaction of the external field H with the microscopic field B, plus the Maxwell stress tensor of the microscopic field B. If $H = 0$, $B = 4\pi\mathcal{M}$ and $\mathcal{T}_{ij} = P_0\delta_{ij} + \mathcal{S}(B)_{ij}$. In this case the magnetic field is kept self-consistently.

At this point we want to refer to the recent paper by Khalilov [4] where a similar problem to the present one is studied. However, the expression for the stress tensor in a medium is taken as given only by the linear approximation of the Minkowski tensor term. This approach is not justified in the relativistic case. Further, the Maxwell tensor of the microscopic field B and the isotropic $P_0\delta_{ij}$ terms are omitted. This leads the author to wrongly conclude that the collapse occurs like in the classical case (see below), that is, along the **B** field, in contradiction to our present results and those of [1]-[2]. A consistent approach, as we have followed here, and discussed in the accompanying paper [21], leads to opposite results compared to those of Ref. [4] in what concerns the spatial direction of collapse, although the fundamental issue regarding the collapse of the neutron star remains taking place in both theories.

Note, in addition, that Refs. [6] and [7], where the problem was also studied, did not take into account the dynamical effects of the strong (and super strong) magnetic fields supposed to exist in the core of canonical NSs, see for instance Refs. [1] [22]-[23]. Thence, it is the contention of this paper to address this open issue. The novel results obtained here point out to the occurrence of new physics and processes in the (relativistic) astrophysics of compact objects that were not manifest in previous papers.

4 Classical vs. Quantum Collapses

In Ref.[1] we found that a relativistic degenerate electron gas placed in a strong external magnetic field B is confined to a finite set of Landau quantum states. As the field is increased the maximum Landau quantum number is decreased favoring the arising of either a paramagnetic or a ferromagnetic response through a positive magnetization $\mathcal{M}$, up to the case in which only the ground state is occupied. The gas then becomes topologically one dimensional, and in consequence the pressure transverse to the field vanishes for fields $B = \Omega/\mathcal{M}$ [1]. Thus, the electron gas becomes unstable due to the decrease of the transverse pressure for fields strong enough, and the outcome is a collapse.

For neutrons the magnetization is always positive (see arguments below) and nonlinear, what leads to a sort of ferromagnetic behavior. For fields strong enough the pressure transverse to the field, $P_\perp = -\Omega - B\mathcal{M}$, is considerably decreased and may vanish. If we assume that extremely magnetized NSs, as the Duncan and Thompson magnetars[3], have fields $H \sim 10^{15}$ G, and that inside the star B increases by following a dipole law $B(r') = B_{surf}/r^3$, we expect near its surface magnetic fields ranging from $10^{16} - 10^{17}$G[3] and [19] up to values of order 10^{20}G in its core [20], where the field is maintained self-consistently, i. e., $H = 0$. For fields of this order of magnitude super dense matter composed of neutral particles having a magnetic moment may undergo a transverse collapse since $P_\perp$ vanishes. As discussed below, the emerging physics seems to ban the possibility of magnetar formation. The outcome of such a collapse might be a compact star endowed with canonical magnetic field, as discussed below.

In the classical case in which the response of the medium is due to the Lenz law, the magnetization is opposite to the external field H and it may happen that $\mathcal{M} < 0$. This also occurs in the diamagnetic case. Then $H > B$ and $P_\perp > P_3$. Note that the opposite occurs in some permeable materials where $\mathcal{M} > 0$ and $H = B - 4\pi\mathcal{M}$ is small in comparison to either $\mathcal{M}$ and B; this is due to ferromagnetic effects which have quantum origin, as in the neutron gas.

In the case of a classical magnetized gas, as $P_\perp > P_3$, this leads to the Earth-like oblatening effect described above. But opposite to this, for the critical quantum configuration of the NS gas the coupling of the spin magnetic dipole with the magnetic field $\vec{B}$ plays the main role, and $\mathcal{M} > 0$ (see Eq.(21) below and the subsequent discussion where this is shown explicitly), leading to ferromagnetic effects. The situation then is reversed and $P_\perp$ is smaller than P_3 in the amount $B\mathcal{M}$ and it vanishes for $\frac{1}{8\pi}B^2 = \frac{1}{4\pi}\int BdH + P_0$ leading, conversely, to a prolate configuration.

In classical electrodynamics [10] it is suggested that the total pressure is given by the sum of the Maxwell stress tensor $\mathcal{S}_{\mu\nu}$ plus an isotropic pressure (P_0) term. In the case of a constant magnetic field parallel to the x_3 axis, the total pressure tensor reads $\mathcal{T}_{ij} = P_0\delta_{ij} + \mathcal{S}_{ij}$ or $P_3 = P_0 - B^2/8\pi$ and $P_\perp = P_0 + B^2/8\pi$.

As pointed out before, the body deforms under the action of these anisotropic pressures. If the longitudinal pressure decreases, the body flattens along the magnetic field [24]. Thus, in this pure classical case, for the extreme limit of flattening, $P_3 = 0$ and $P_\perp = P_0 + B^2/8\pi$, and the body would collapse as a disk or a ring perpendicular to the field. Starting from general relativistic considerations it has been reported [22] the existence of a maximum magnetic field for having stationary configurations of NSs. (We interprete that result as indicating the occurrence of a classical collapse for fields larger than the maximum). This field induces a toroidal configuration, which is topologically equivalent to a ring. In the quantum case, for degenerate fermions, as $\mathcal{M} > 0$, it is $P_\perp = -\Omega - B\mathcal{M}$ which is decreased by increasing B. As the NS is in equilibrium under the balance of neutron and "gravitational" pressures, the body stretches along the direction of the magnetic field. Thus, for any density there are values of the field B high enough such that these pressures cannot counterbalance each other leading to a collapse perpendicular to the field for $P_\perp = 0$. This collapse would leave as a remnant a nucleons plus a Bose-Einstein-like condensate, a hybrid or strange star (The reader is addressed to the interesting references: "What if pulsars are born as strange stars?", by R. Xu, B. Zhang and G. Qiao, Astropart. Phys. 15, 101 (2001), and "PSR 0943+10: A bare strange star?", by the same authors and published in Astrophys. J. 552, L109 (1999)), with canonical magnetic field [25], or a distorted ("cigar-like") black hole.

Our previous considerations are approximation-independent. In order to discuss an specific model, we shall start by computing the free particle spectrum for neutral particles in a magnetic field.

5 The Neutron Gas in a Magnetic Field

For free neutrons in a magnetic field B we have the Dirac equation for neutral particles with anomalous magnetic moment [14]

$$(\gamma_\mu\partial_\mu + m + iq\sigma_{\mu\lambda}F_{\lambda\mu})\psi = 0, \tag{11}$$

where $\sigma_{\mu\lambda} = \frac{1}{2}(\gamma_\mu\gamma_\lambda - \gamma_\lambda\gamma_\mu)$ is the spin tensor, and $F_{\lambda\mu}$ is the electromagnetic field tensor describing B. By solving this equation we get the eigenvalues [14]

$$E_n(p, B, \eta) = \sqrt{p_3^2 + (\sqrt{p_\perp^2 + m_n^2} + \eta q B)^2}, \tag{12}$$

where the subscript n refers to the Landau levels, p_3, $p_\perp$ are respectively the momentum components along and perpendicular to the magnetic field B, m_n is the neutron mass, $q = 1.91\ M_n$ (M_n is the nuclear magneton), $\eta = 1, -1$ are the σ_3 eigenvalues corresponding to the two orientations of the magnetic moment (parallel and antiparallel) with regard to the field B. Here we make the assumption that the magnetic moment remains constant for large magnetic fields. Actually, this is not so. Radiative corrections change its value as a function of B. This problem is, however, beyond the scope of the present approach.

From (12) we see that although the Hamiltonian is linear in B, the eigenvalues depend on B nonlinearly. This makes the relativistic thermodynamic and electromagnetic properties of the system of neutrons very different from the nonrelativistic case. For instance, all thermodynamic quantities, Ω, $\mathcal{M}$ (and in consequence, $H = B - 4\pi\mathcal{M}$), and N are also nonlinear functions of B. The expression (12) also shows manifestly the change of spherical to axial symmetry with regard to momentum components. *This anisotropy in the dynamics is expected to be reflected in an anisotropy in the thermodynamic properties of the system, as it is expressed by the difference between the transverse and longitudinal pressures discussed earlier on an approximation-independent basis.*

The partition function $\mathcal{Z} = \mathrm{Tr}(\rho)$ is obtained from the density matrix describing the model $\rho = e^{-\beta \int d^3x(\mathcal{H}(\mathbf{x}) - \sum \mu_r N_r)}$ [26]. Here μ_j ($j = 1, 2, 3$) are the chemical potentials associated with the lepton, baryon and electric charge conservation, so that $\mu_n = \mu_2$, $\mu_p = \mu_2 + \mu_3, \mu_e = \mu_1 + \mu_3$ and $\mu_\upsilon = \mu_1$. The thermodynamical potential can be written as $\Omega = -\beta^{-1} ln \mathcal{Z}$. Usually the eigenvalues of $\mathcal{H}$ contain the contribution from neutrons, protons, electrons and some meson species, and the densities are $N_r = -\partial\Omega/\partial\mu_r$, where $r = n, p, e...$ We name $\Omega = \sum_r \Omega_r$, $\mathcal{M} = \sum_r \mathcal{M}_r$ the total thermodynamical potential and magnetization, respectively.

A standard procedure is to work in the mean field approximation in which the meson fields σ, ρ, ω are taken as constant, as done in Refs.[20], [23], through which the mass spectrum of baryons is corrected and strong repulsive interactions between them is found. However, for simplicity we will keep the spectra in the tree approximation to obtain the one-loop approximation for Ω, and neglect the statistical contribution from meson terms in Ω as compared with those of fermions (for them $m_i\beta \sim 10^3 - 10^4$) except for fields $B \leq B_{c\rho} = m_\rho^2/e \approx 10^{20}$ G, since the contribution of the ρ vector meson condensate to $\mathcal{M}$ becomes relevant, and in analogy with $W^\pm$-s[1], leads to a self-consistent spontaneous magnetization $B = 4\pi\mathcal{M} = 2\pi e N_\rho \sqrt{m_\rho^2 - eB}$, where N_ρ is the condensate density. However, for such fields the magnetic pressure: $\int \mathcal{M} dB - \mathcal{M}B = -B^2/8\pi$ overwhelms the kinetic pressure term P_0 (of order 10^{36} dynes cm^{-2}) leading to $P_\perp < 0$, and the star is definitely unstable: it collapses. This mechanism is valid for other quasi-particle vector boson condensates, as di-quarks, which may be formed in the medium even for smaller values of the field B [25, 13].

6 Thermodynamical Potential of a Neutron Gas

The Green functions method [27] leads to a general expression for the relativistic thermodynamical potential. At the one-loop level, where no radiative corrections are considered, it is a generalization of the usual nonrelativistic formula because of the fact that antiparticles must also be included. Particles and antiparticles contribute with chemical potentials of opposite sign, leading to sums or integrals over the quantum numbers involved, of terms containing the product of the logarithms of $(e^{-(\mu_n \pm E_n)} + 1)$. In this case, when no external fields are present, a divergent term accounting for the vacuum energy appears which is usually subtracted [27]. In presence of an external field, however, a term accounting for the vacuum contribution, must also be included: The Euler-Heisenberg energy of vacuum in an external field [11, 15, 26, 28]. One can obtain an expression for the thermodynamic potential of the neutron gas in the one-loop approximation as $\Omega_n = \Omega_{sn} + \Omega_{Vn}$, with

$$\Omega_{sn} = -\frac{1}{4\pi^2\beta} \sum_{\eta=1,-1} \int_0^\infty p_\perp dp_\perp dp_3 \ln\left[f^+(\mu_n,\beta) f^-(\mu_n,\beta)\right], \tag{13}$$

where $f^{\pm}(\mu_n,\beta) = (1 + e^{-(E_n \mp \mu_n)\beta})$ accounts, respectively, for the contribution of particles and antiparticles. The expression for the vacuum term reads thus

$$\Omega_{Vn} = \frac{1}{4\pi^2\beta} \sum_{\eta=1,-1} \int_0^\infty p_\perp dp_\perp dp_3 E_n \tag{14}$$

which is divergent. In the Appendix we will show how to regularize this expression, and how to obtain the analog to the Euler-Heisenberg energy of vacuum due to the neutrons contribution.

After integrating by parts in (13) its evaluation becomes easier. The Fermi distributions, which arise by differentiating $f^{\pm}$ with respect to p_3, are $n^{\pm} = 1/(1 + e^{(E_n \mp \mu_n)\beta})$. In the degenerate case this expression reduces in $n^- = \theta(\mu - E_n)$ and $n^+ = 0$, since in that case only particles contribute to Ω. The resulting expression splits itself in two terms where the integrals are bounded by the Fermi surfaces $\mu - E_n(\eta = \pm 1) = 0$. These surfaces have axial symmetry, and thus, the Fermi momentum is not a definite number, given only in terms of μ_n and m_n, but on the opposite, it has infinite values. Thence, we have

$$\Omega_{sn} = -\frac{1}{4\pi^2} \sum_{\eta=1,-1} \int_0^\infty p_\perp dp_\perp \frac{p_3 dp_3}{E_n} \theta\left(\mu - E_n(\eta)\right). \tag{15}$$

The θ function bound these integrals in the intervals $-p_{3F} \leq p_3 \leq p_{3F}$, where $p_{3F} = \sqrt{\mu^2 - (\sqrt{p_\perp^2 + m_n^2} + \eta y)}$ and $0 \leq p_\perp \leq \sqrt{(\mu - \eta y)^2 - m_n^2}$. After some transformations, it yields

$$\Omega_{sn} = -\Omega_0 \sum_{\eta=1,-1} \left[\frac{x f^3}{12} + \frac{(1+\eta y)(5\eta y - 3)x f}{24}\right.$$

$$+ \frac{(1+\eta y)^3(3-\eta y)}{24}L - \frac{\eta y x^3}{6}s\Bigg], \tag{16}$$

where $x = \mu_n/m_n$, ($m_n(x-1)$ is the usual Fermi energy), and $y = qB/m$. We define the functions $f \equiv f(x, \eta y) = \sqrt{x^2-(1+\eta y)^2}$, $s \equiv s(x, \eta y) = (\pi/2 - \sin^{-1}(1+\eta y)/x)$, $L \equiv L(x, \eta y) = \ln(x + f(x, \eta y))/(1+\eta y)$. The functions $f =< p_F > /m_n$, where $< p_F >= \sqrt{\mu_n^2 - (m_n + \eta qB)^2}$ and we name them the average Fermi momenta for $\eta = \pm 1$. We see that $m_n - \eta qB$ behaves formally as a two-valued magnetic mass.

In the zero field limit one gets $P_0 = -\Omega_n(y=0)$, where

$$\Omega_{sn}(y=0) = -\Omega_0\left[\frac{x f_0^3}{12} - \frac{x f_0}{8} + \frac{1}{8}L_0\right], \tag{17}$$

where $f_0 = \sqrt{x^2-1}$ is the relative Fermi momentum p_F/m, and $L_0 = \ln(x + f_0)$. The neutron vacuum term (see Appendix) has an Euler-Heisenberg-like form as

$$\begin{aligned}\Omega_{Vn} &= \frac{1}{4\pi^2}\int_0^\infty dy y^{-3} e^{-(m_n^2+q^2B^2)y}[\cosh(qBmy) - 1 - (qBmy)^2/2!] \\ &+ \frac{qB}{2\pi^2}\int_0^\infty dy y^{-2}\int_0^\infty dw e^{-[(w+m_n)^2+q^2B^2]y}[\sinh(2qB(w+m_n)y) \\ &- (2qB(w+m_n)y) + (2qB(w+m_n)y)^3/3!]\end{aligned} \tag{18}$$

It can be shown (see Appendix) that the more significant term in (18) is the first one, which for fields of order 10^{17} G leads to $\Omega_{Vn} \sim 10^{30}$ erg·cm^{-3} and is negligible small as compared with Ω_{sn} up to $B \sim 10^{18}$ G. Thus, we neglect it in a first approximation in what follows. We must point out, however, that since neutrons have a quark structure, a more fundamental quantity would be the vacuum quark contribution, whose order of magnitude is expected to be near Ω_{Vn}. Apart from this note, it should be emphasized that the role of vacuum cannot be ignored for fields greater than 10^{18} G.

From $N_n = \partial\Omega/\partial\mu_n$ one gets

$$N_n = N_0\sum_{\eta=1,-1}\left[\frac{f^3}{3} + \frac{\eta y(1+\eta y)f)}{2} - \frac{\eta y x^2}{2}s\right], \tag{19}$$

In the limit $B = 0$ Eq.(30) reproduces the usual density of a relativistic Fermi gas at zero temperature, $N_n = N_0 f_0^3/3$.

Having an equation relating the chemical potentials, and demanding conservation of both baryonic number $N_n + N_p = N_B$ and electric charge $N_p + N_e = 0$, in principle one may solve exactly the problem in terms of the external field as a parameter. Nonetheless, we shall focus our discussion on the properties of the equation of state. Note in passing that our expressions for the spectra and densities of neutrons and protons are similar to those of Ref. [29] of a neutron gas in a magnetic field, but we get different equations of state.

Finally, for the magnetization, given as $\mathcal{M}_n = -\partial\Omega_n/\partial B$, we have

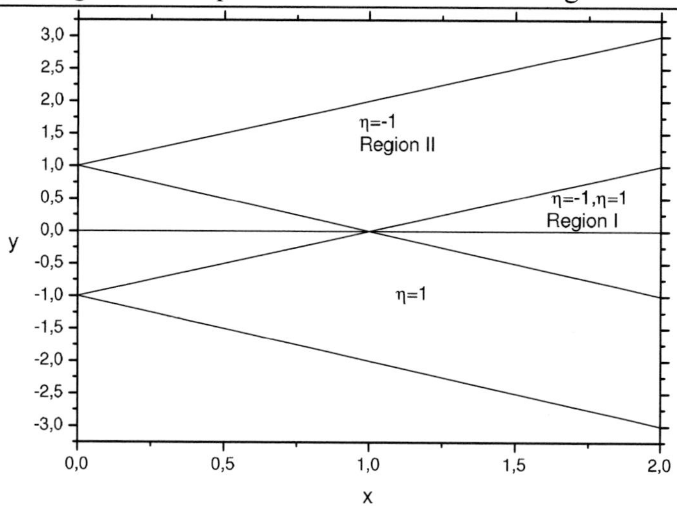

Figure 1: This plot shows the region in the plane where the neutron magnetic moments are oriented parallel or antiparallel to the magnetic field $\vec{B}$. Special attention should be given to regions I and II, where the solutions discussed in the text are valid.

$$\begin{aligned} \mathcal{M}_n &= -\mathcal{M}_0 \sum_{\eta=1,-1} \eta \left[\frac{(1-2\eta y)xf}{6} \right. \\ &- \left. \frac{(1+\eta y)^2(1-\eta y/2)}{3} L + \frac{x^3}{6} s \right] \end{aligned} \tag{20}$$

where $N_0 = m_n^3/4\pi^2 \sim 2.04 \times 10^{39}$, $\Omega_0 = N_0 m_n \sim 3.0 \times 10^{36}$, and $\mathcal{M}_0 = N_0 q \sim 2.92 \times 10^{16}$ and one can write $\mathcal{M}_n = \mathcal{M}_n^+(\eta = -1) - \mathcal{M}_n^+(\eta = +1)$, and obviously, $\mathcal{M}_n \geq 0$.

We confirmed by explicit calculation that $\mathcal{M}$ is a nonlinear function of B and, in this sense, the magnetic response is ferromagnetic. A fully ferromagnetic response demands to include also the spin-spin coupling contribution. We discuss briefly this point below.

To see why the magnetization is always positive for the neutron gas note that the magnetic susceptibility $\chi = \partial \mathcal{M}_n/\partial B$ can be easily obtained as

$$\chi = \frac{q\mathcal{M}_0}{2m_n} \sum_{\eta=\pm 1} [xf + (1+\eta y)^2 L], \tag{21}$$

which for $x > 1$, $y \leq 1$, and f real and positive, it is $\chi > 0$. This means that $\mathcal{M}_n$ is an increasing function of B (or y) under such conditions. As $\mathcal{M}_n(y = 0) = 0$ and $\mathcal{M}_n(y = 1) = 2\mathcal{M}_0(1 + \pi) > 0$, this means that $\mathcal{M}_n > 0$ in the region between these two points, which is the one of interest for us (the region I discussed below and showed in Fig. 2).

The fact that we are summing over the magnetic moments oriented parallel ($\eta = -1$) and antiparallel ($\eta = +1$) to B is similar to the well known Pauli paramagnetism in nonrelativistic quantum statistics. We may consider each term $\eta = \pm 1$ as representing a phase of the system. One can draw in the x, y plane, for both $x > 0$, $y > 0$, two regions limited by the lines $x - y = \pm 1$ and $x + y = 1$ (see Figure 1). The region I is bounded above by the line $x = y + 1$, on which the contribution from $f(\eta = +1)$, and all other terms containing $\eta = +1$ vanish: all magnetic moments are aligned parallel to B. Below, such region is limited by $y = 0$. For points in the region $x > y + 1$ and $y > 0$ both terms containing $\eta = -1, +1$ are nonzero and real. Thus, the quantities Ω, N_n, and $\mathcal{M}$ are real and although most neutrons have their magnetic moments along B, there are some amount of them having their magnetic moments antiparallel to B. In this region, near the line $x = y + 1$, is located the curve $P_\perp = 0$, which we will discuss below.

In the region II limited by $x \geq 1-y$, $y+1 > x \geq y-1$ only the term $\eta = -1$ is real and the term $\eta = +1$ becomes purely imaginary, and all neutrons would have their magnetic moments parallel to B. The term $\eta = -1$ contribute to Ω, N_n, and $\mathcal{M}$. For $x < y - 1$, there are no physical solutions. For $y = 0$, the magnetic field, and in consequence, the magnetization $\mathcal{M}$ vanish. For $y < 0$ (magnetic field along the negative y axis) we have the reverse situation to the one described above (See Figure 1).

With regards to the background of electrons and protons, we remind that the case of the electron gas was discussed in ([1]). We expect that for magnetic fields of order $B_{ce} = m_e^2/e \sim 10^{13}$ G and densities around 10^{30} cm^{-3} all electrons are in the Landau ground state, and the system show the instability which arises from the vanishing of the transverse pressure. For densities and magnetic fields above these critical values, the stability of the electron gas is doubtful. Recently Ghosh, Mandal and Chakrabarty [5] has pointed out strong arguments against the equilibrium of an electron gas under β decay in strongly magnetized neutron stars. The situation may be modified if radiative corrections are taken into account: if $B_{ce} = m_e^2/e \sim 10^{13}$G is the usual QED critical field, for larger fields of order $B'_{ce} \sim 4\pi/\alpha B_{ce} \sim 10^{16}$G the contribution of the electron anomalous magnetic moment becomes significant and the problem cannot be satisfactory treated at the tree level. However, starting from the results of [1],[5], one concludes that the electron gas is hardly in equilibrium for fields beyond B_{ce}. One possibility is the bosonization of the electron system, as has been recently suggested [30]. This may be accomplished through the increase of the spin-spin interaction, leading to the formation of parallel spin electron pairs, equivalent to charged vector bosons.

If we include both the normal and the anomalous magnetic moments for protons, one can give a formula for their spectrum in the external field B as[14]:

$$E_p = \sqrt{p_3^2 + (\sqrt{2eBn + m_p^2} + \eta q_p B)^2}, \tag{22}$$

$q_p = 2.79 M_n$. For neutrons, the critical field at which the coupling energy of its magnetic moment equals the rest energy is $B_{cn} = 1.57 \times 10^{20}$ G. For protons it is $B_{cp} = 2.29 \times 10^{20}$ G. By defining $x_p = \mu_p/m_p$, $y_p = q_p/m_p$, $b = 2e/m_p^2$, then $y_p = 2.79e/2m_p^2$. We name also $g \equiv g(x_p, B, n) = \sqrt{x_p^2 - h(B,n)^2}$ and $h \equiv h(B,n) = (\sqrt{bBn+1} + \eta y_p B)$. Thus for the proton thermodynamical potential we get

$$\Omega_p = -\frac{eBm_p^2}{4\pi^2} \sum_n \sum_{\pm\eta} \left[x_p g - h^2 \ln(x_p + g)/h \right], \tag{23}$$

and for its density

$$N_p = \frac{eBm_p}{2\pi^2} \sum_n \sum_{\pm\eta} g(x_p, B, n) \,, \tag{24}$$

while the magnetization is given by

$$\begin{aligned} \mathcal{M}_p &= \frac{em_p^2}{4\pi^2} \sum_n \sum_{\pm\eta} \Big\{ x_p g - \Big[h^2 + (\eta y_p \\ &+ (bn/2\sqrt{bBn+1})) \Big] \times \ln\left(x_p + g\right)/h \Big\} , \end{aligned} \tag{25}$$

where the coefficients of these formulae are $N_{0p} = em_pB/2\pi^2 \sim 4.06 \times 10^{19}B$, $\Omega_{0p} = N_{0p}m_pB \sim 6.1 \times 10^{16}B$, and $\mathcal{M}_{0p} = N_{0p}m_p = \Omega_0/B$. The maximum occupied Landau quantum number n may be given as $n_{max} = (x_p - \eta y_p B)^2 - 1/bB$. For $B \ll B_{cp}$, so that $y_pB \ll 1$, and $x_p \geq 1$, one can take approximately $n_{max} \sim (x_p^2 - 1)/bB$, and for fields large enough $n_{max} = 0$. We expect that from the equation $\mu_n = \mu_p + \mu_e$, then $x_p \sim x_n$, the previous expression for the proton density N_p decreases with increasing B, favoring the inverse beta decay. For fields $B \sim m_p/q_p$ and $x_p \gg 1$, $n_{max} \geq 1$, and thus large Landau numbers are again occupied. However, for $x_n, x_p \geq 1$, being both quantities of the same order of magnitude, from the comparison of $\Omega_{0p}, N_{0p}, \mathcal{M}_{0p}$ with $\Omega_0, N_0, \mathcal{M}_0$ we conclude that for fields below 10^{19} G, the dominant longitudinal pressure, density and magnetization comes from the neutron gas.

7 Condition for Zero Transverse Pressure and Collapse

In the electron gas case [1], the vanishing of the transverse momentum can be guessed from the spectrum when all the system is confined to the Landau ground state. The spectrum corresponds to a purely one-dimensional system moving along the external field, and the transverse Fermi momentum is zero. In a similar way, in the neutron gas case we observe that the threshold of zero transverse pressure $P_\perp = 0$ can be figured out from the spectrum described by Eq.(12), since the contribution from the $\eta = -1$ term is dominant (observe that the term with $\eta = +1$ contributes with a negative term to $\mathcal{M}$). We shall consider on the Fermi surface for $\eta = -1$ the quantity

$$p_{F\perp eff}^2 = \mu_n^2 - p_{F3}^2 - m_n^2 = \left(\sqrt{p_{F\perp}^2 + m_n^2} - qB\right)^2 - m_n^2 \tag{26}$$

which we name the effective squared Fermi transverse momentum. If $B \ll 10^{20}$ G, then $q^2B^2 \ll 2eBm_n$, the vanishing of $p_{F\perp eff}$ is guaranteed if

$$p_{F\perp} \sim \sqrt{2qBm_n}. \tag{27}$$

The resulting Fermi surface would be equivalent to that for one-dimensional motion, parallel to B, i. e., for particles having energy $E_n \simeq \sqrt{p_3^2 + m_n^2}$, and in consequence the transverse momentum (and pressure) vanishes. Notice from (27) that for $p_{F\perp}/m_n \sim 10^{-1.5}$

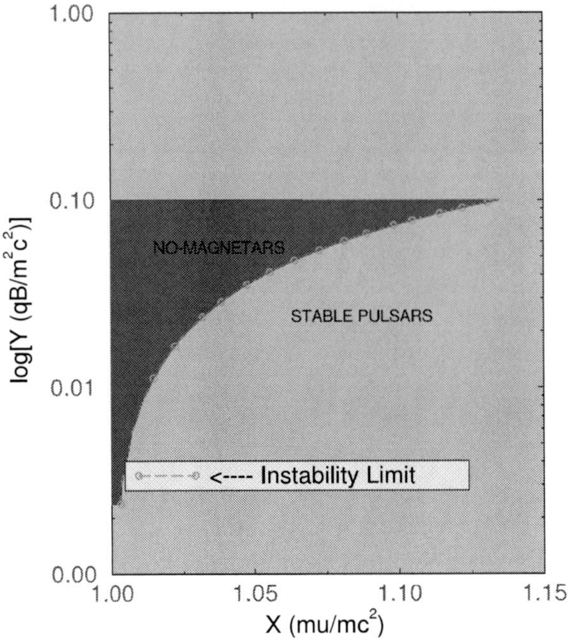

Figure 2: The instability condition: $P_\perp = 0$ in terms of the variables x, y. A nascent strongly magnetized neutron star having a configuration such that its dynamical stage would be represented by a point above the central curve in this plot would be unstable to transversal collapse, since $P_\perp \leq 0$ there.

one has $qB/m_n = y \sim 10^{-3}$, which means fields of order 10^{17}G. A more accurate result is obtained, however, from the condition: $\mathcal{T}_\perp = 0$.

In Figure 2 we have drawn the equation $P_{\perp n} = -\Omega - B\mathcal{M} = 0$ in terms of the variables x, y. We observe that there is a continuous range of values of x, starting from $x = 1.005$ to $x = 1.125$ and from $y = 0.001$ to $y = 0.1$, for which the collapse takes place. The latter values of y means fields in the interval $B \sim 10^{17}$ to 10^{19} G. To these ranges of x, y corresponds a continuum range of densities, from $10^{-2}N_0$ (10^{12} g/cm^3) onwards. The transverse compression of the whole mass of the star due to flux conservation leads to an increase of B and the mechanism of collapse is enhanced.

8 The Spin-Spin Coupling Contribution

Although we used previously the term ferromagnetic to denote the non-linear response of the medium to the external field, what we have considered actually in our previous calculations is the occurrence of relativistic Pauli's paramagnetism. We have not fully considered the spin-spin coupling, which would lead more definitely to Heisenberg ferromagnetism, and is physically reasonable to expect its appearance in nuclear matter for densities high enough. This problem has been studied in [31] via the interaction through axial vector and tensor exchange channels. These authors show that if the interactions are strong enough and differ in sign, the system loses the spherical symmetry due to a mechanism independent of the one discussed in the present paper. In the case of a quark liquid [32] the problem has been studied under one gluon exchange interaction, and conditions for ferromagnetism may arise.

One expects that if Heisenberg ferromagnetism is to appear, it would increase the magnetization to $\mathcal{M}_s = \kappa\mathcal{M}$, where κ is the internal field parameter. If $\kappa >> 1$ our previous estimates for the vanishing of the transverse pressure might be largely exceeded, with the arising of a new spontaneous magnetization $\mathcal{M}_s \sim \kappa\mathcal{M} = B/4\pi$. This would mean that the magnetic field B could be kept self-consistently and our previous calculations would be a lower bound of the fully ferromagnetic case. The vanishing of $P_\perp$ is expected to occur surely at values of B smaller than those depicted in Figures 2 and 3.

We will consider in what follows an heuristic model of ferromagnet resulting from the assumption of neutron pair coupling, leading to a neutral vector boson bearing a magnetic moment. This would lead to an approximate model for ferromagnetic coupling among neutrons, which is expected to arise in presence of extremely strong magnetic fields. Because of this special configuration one expects that the resulting Bose particle to have a mass of order twice the fermion masses originating it. The idea of vector pairing (i.e. parallel spin-pairing in place of scalar or antiparallel pairing) is justified if we start from the essential property that the neutron ground state corresponds to spin alignment along the external field, and for fields large enough the difference between parallel and antiparallel pairing is a significant fraction of the pair energy. It is expected then that the spin parallel is a more energetically favorable coupling than the antiparallel or scalar one. Thus, the superfluid, which is the physical realization of the vector condensate, is also a ferromagnet. For charged particles the situation is similar, and for electrons, this has lead to the suggestion of re-entrant superconductivity, for some condensed matter systems in magnetic fields of order 10^6 G. In this system a large population of parallel spin electrons is expected to occur in the Landau ground state $n = 0$ (See Boebinger and Passner [33] and references quoted therein).

We start by considering a relativistic neutral boson gas of particles with nonzero magnetic moment. A variety of physical arguments allow for the bosons to be described as a neutral vector gauge field, even if it is the result of a pairing of fermions. Its spectrum can be derived from the poles of the inverse propagator for a massive gauge neutral boson that couples to an external magnetic field through the coupling constant $g = Mq$, where it is expected that due to interaction it is $M = M(B)$ and we will assume $M \sim 2m_n$ and q has the dimensions of charge. The inverse propagator reads

$$D^{-1}(k_\mu, F_{\mu\nu}, M) = (k^2 + M^2)\delta_{\mu\nu} - k_\mu k_\nu - igF_{\mu\nu}, \tag{28}$$

where $F_{\mu\nu} = B(\delta_{\mu 1}\delta_{\nu 2} - \delta_{\mu 2}\delta_{\nu 1})$ is the electromagnetic field tensor describing the field B, which is pointing along the 3-rd axis in this our treatment. It is easy to find the spectrum from $DetD^{-1}(k_\mu, F_{\mu\nu}, M) = 0$ as

$$E(p, B) = \mp\sqrt{p_3^2 + p_\perp^2 + M^2 + \eta(\sqrt{p_\perp^2 + M^2})qB}, \tag{29}$$

where $\eta = \mp 1$ and the minus and plus signs in front of the square root indicate particles aligned respectively along and opposite to the magnetic field B. The ground state corresponds to the case $\eta = -1$. The magnetic moment of the particles is in this case defined as:

$m = -\frac{\partial E(0,B)}{\partial B} = \frac{q}{\sqrt{M^2 - MqB}}$.

Starting from Eq.(29), one can write an expression for the thermodynamical potential $\Omega = -T \ln \mathcal{Z}$, where $\mathcal{Z}$ is the partition function. Once Ω is known, if μ is the chemical potential, the density is given by the formula $N = -\frac{\partial \Omega}{\partial \mu}$. Its expression reads

$$N = \frac{2\pi}{\hbar^3} \Sigma_\eta \int_{-\infty}^{\infty} p_\perp dp_\perp dp_3 [\frac{1}{e^{(E(\eta)-\mu)/T} - 1} - \frac{1}{e^{(E(\eta)+\mu)/T} - 1}], \tag{30}$$

One can show that for fixed N, μ is a decreasing function of T. Thus, by decreasing T (or by increasing N) μ increases, and reaches its maximum value at the ground state energy (magnetic moment up) $\mu = \sqrt{M^2 - MqB}$, at some critical temperature T_c. In other words, the neutral particle temperature Green's function has an infrared pole at $p_4 = i\omega = \pm i(E(0,B) - \mu) = i(\sqrt{M^2 - MqB} - \mu) = 0$.

As in the usual theory of Bose-Einstein condensation, for temperatures $T < T_c$, the integral $N(T) < N$ gives the number of excited particles, the condensate being $N_0 = N - N(T) = N(1 - (T/T_c)^{3/2}$. Obviously, due to the magnetic field the system is subject to strong magnetostrictive forces expressed in anisotropic pressures [34]-[9]: they are different along and perpendicular to **B**. The anisotropic pressures for the gas of magnetized particles were obtained for charged particles in previous papers [1], and for neutral particles in [35]. The pressure along the field is $P_3 = -\Omega$, and perpendicular to it is $P_\perp = -\Omega - B\mathcal{M}$, where $\mathcal{M}$ is the magnetization. Such an anisotropy can be compensated by deforming the body, but in some limits (discussed in Refs. [1][35] and extended here to BECs), it leads to a collapse of the system of particles.

One can estimate that near the condition for condensation, by calling $\bar{M} = \sqrt{M^2 - MqB}$, the density, Eq.(30), of excited particles is approximately given by

$$N \sim T\sqrt{\bar{M}^2 - \mu^2}/4\pi, \tag{31}$$

and decreases as $\mu \to \bar{M}$. The isotropic pressure decreases more strongly, as $P_0 = -\Omega \sim \frac{T}{6\pi}(\bar{M}^2 - \mu^2)^{3/2}$. In other words, the condensate does not contribute to the thermodynamical potential, or what is the same, to the longitudinal pressure. The thermodynamic potential of the condensate is thus zero, but its (rest) energy is non-zero, since $U_0 = \mu N = \bar{M}N$. The condensate contributes, however, negatively to the transverse pressure, as we shall see below.

9 Magnetic Behavior of the Bosonized Gas

In general, it is expected to have a background of excited bosons and fermions, and thus, the total transverse pressure have contributions of both fermions and excited bosons. Meanwhile, the magnetization contains in addition the contribution from the boson condensate, which may be very large. For bosons, after the phase transition to a condensate, the magnetization must be written as $\mathcal{M}_b = -\partial\Omega/\partial B + N_0 m$, where the second term contains the contribution to the magnetization due to the condensate energy, $N_0 \partial \bar{M}/\partial B = N_0 m$. In our present case, the total transverse pressure is then $P_{t\perp} = -(\Omega_b + \Omega_f) - B(\mathcal{M}_b + \mathcal{M}_f)$ [1, 35], where t refers to total, and b and f refer, respectively, to bosons and to fermions. If there is a large amount of bosonization and most of the particles are in the condensate, i.e., when $-(\Omega_b+\Omega_f)$ is equal or smaller than the negative pressure term $-B(\mathcal{M}_b+\mathcal{M}_f)$, either the total pressure perpendicular to the field vanishes or becomes negative and the system collapses. We consider two cases: the relativistic neutron star, and the non-relativistic gas of few millions of atoms, that is a BEC.

For a neutron star where there is neutrons' spin-pairing parallel to B, which leads to an effective spin one boson particle as the one described above, having an effective mass $2m_n$. Thus, even assuming temperatures of $\sim 10^8$K, since $2m_n/T \sim 10^5$, the system must be considered as highly degenerate, that is, below the critical temperature ($T_c \geq 10^{11}$ K), and the density of excited particles is negligible small as compared with the condensate density. Also, the Bose-Einstein condensate would lead to a more energetically favorable state than the fermion gas, as is seen by starting from the general expression for the internal energy density, $U = \mu N + TS + \Omega$. As for the condensate it is $\Omega = 0$ and $S = -\partial\Omega/\partial T = 0$, its internal energy is $U_b \sim \mu_b N_b$, where $\mu_b \leq 2M_f$ and $N_b \leq N_f/2$, whereas for the Fermi gas it is $U_f \sim \mu N_f + \Omega_f$, where Ω_f is positive and of order μN_f, and $\mu_f > M_f$. Thus, $U_b < U_f$. As pointed out above, the Bose condensate bears the properties of a superfluid and also of a ferromagnet. Therefore, most of the system must be in the condensate state.

If initially a strong magnetic field is created locally in the star (say, of order $B_i \sim 10^{12-13}$ G, as inherited from the implosion of the progenitor (a supergiant red star) by the mechanism of flux conservation, that field would be amplified by the Bose-Einstein condensation. As the particles in the condensate are described by the spectrum of Eq.(29), then in the limit of zero momentum, and for $\eta = -1$, the condensate becomes polarized. Thence, one can estimate the magnetization as given as $\mathcal{M}_f = N_0 m$ (the thermodynamical potential of the excited particles is obviously zero).

By taking $N_0 \sim 10^{39}$, $M \sim 10^{-24}$, one gets that the system generates a spontaneous magnetization $\mathcal{M}_f \sim 10^{16}$ G, leading to a self-consistent field $B_f = 4\pi\mathcal{M}_f \sim 10^{17}$ G. In this sense, the bosonization we are describing is a simplified version of a more realistic superfluid or ferromagnetic coupling in which particles interact usually with several neighbors, whereas we are assuming in our bosonization model that it results from two-particle vector pairing. Thus, as a lower estimate, by calling B_f, B_i the final and initial fields, respectively, we see that (in the present case) it leads to an internal field parameter of order $x = B_f/B_i \sim 10^{3-4}$. Then, the pressure perpendicular to $\mathbf{B}$ would be negative and of order 10^{33-34} erg cm^{-3}, hence driving a transversal collapse to the boson sector. If the density of bosonized neutrons N_0 is of order of the total star neutron density N, it thermodynamical properties predominate over the fermionic components, leading to a unavoidable collapse of the star. A similar argument are valid [36] for the charged sector of the neutron star, which would lead to fields of the same order $\sim 10^{17}G$.

10 Magnetar Formation and Stability

Next we briefly review the basic ideas supporting the theory of magnetars and then show why these hypothetical objects cannot survive after reaching the claimed super strong magnetic fields. We then present prospectives for a hybrid or strange star to appear as a remnant of the quantum magnetic collapse of a NS. According to Duncan and Thompson[3], NSs with very high dipole surface magnetic field strength, $B \sim [10^{14} - 10^{15}]$ G, may form when

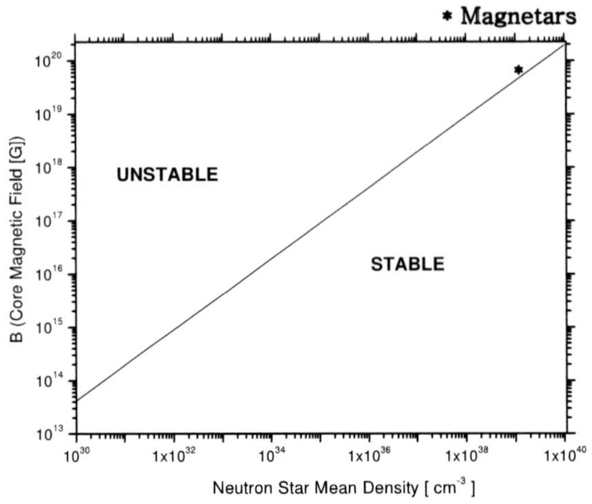

Figure 3: The curve $P_\perp = 0$ in terms of the neutron star mean density N and the surface magnetic field B.

(classical) conditions for a helical dynamo action are efficiently met during the seconds following the core-collapse and bounce in a supernova (SN) explosion[3]. A newly-born NS may undergo vigorous convection during the first 30 s following its formation [37]. If the NS spins (differentially) sufficiently fast ($P \sim 1$ ms) the conditions are created for the $\alpha - \Omega$ dynamo action to be built, which may survive depletion due to turbulent diffusion. Collapse theory, on the other hand, shows that some pre-supernova stellar cores could adquire enough spin so as to rotate near their Keplerian equatorial velocity, the break-up spin frequency: $\Omega_K \geq ([\frac{2}{3}]^3 G_N M/R^3)^{1/2}$, which implies a period $P_K \sim 0.6$ s, after core bounce. Under these conditions, fields as large as [43, 3]

$$B \sim 10^{17} \left(\frac{P}{1\text{ms}} \right) \text{ G}, \tag{32}$$

may be generated as long as the differential rotation is dragged out by the growing magnetic stresses. For this process to efficiently operate the ratio between the spin rate (P) and the convection overturn time scale (τ_{con}), the Rossby number (R_0), should be $R_0 \leq 1$. Duncan and Thompson warned that $R_0 \gg 1$ should induce less effective mean-dynamos[3]). In this case, an ordinary dipole $B_D \sim [10^{12} - 10^{13}]$G may be built by incoherent superposition of small dipoles of characteristic size $\lambda \sim [\frac{1}{3} - 1]$km, and a saturation strength $B_{sat} = (4\pi\rho)^{1/2}\lambda/\tau_{con} \simeq 10^{16}$ G may be reached at the surface during this early evolution of the nascent neutron star. At such fields, the huge rotational energy of a NS spinning at: $\omega_{\rm NS} \geq 1$kHz, is leaked out via *magnetic braking* and an enormous energy is injected into the SN remnant. This process may explain the power emitted by a *plerion*.

As shown above, at the end of the SN core collapse we are left with a rapidly rotating NS endowed with with an extremely strong magnetic field (ESMF) strength and a large matter density $\rho \sim [10^{14} - 10^{12}]$ g cm^{-3}. As illustrated in Figure 1, those are the conditions for the quantum instability to start to dominate the dynamics of the young pulsar. At this stage, the magnetic pressure inwards may overpass the star energy density at its equator and the collapse becomes unavoidable. As the collapse proceeds, higher and higher densities are reached till the point the supranuclear density may reverse the direction of implosion, producing a supernovalike explosion. After which, a hybrid or strange star (SS) may have formed. We explore next this plausible outcome, among the other possibilities quoted above. From that moment, the sound wave generated at the core bounce builds itself into a shock wave traveling through the star at $V_{SW} \sim c/\sqrt{3}$ km s^{-1}. Although the ESMF strength could be quite large as long as the collapse advances, the huge kinetic energy, $E \sim 10^{51-52}$ erg, the mean energy obtained in calculations of energy release in neutron star phase transitions to strange (*twin*) stars [38, 39, 40] and some *prompt shock* supernovae ($E_{p-s} \sim 10^{51}$ erg) [37], carried away by the shock wave drives a kind of supernova explosion inasmuch as in the quark nova model and similar scenarios [40]. Such a huge *ram* pressure may counterbalance the magnetic pressure, and even surpass it, i. e.,

$$\rho_{eject} V_{SW}^2 \geq \frac{B^2}{8\pi\mu_0}\left(\frac{R}{r_A}\right)^6, \tag{33}$$

at a location from the star center equivalent to the Alfvén radius of the magnetar

$$r_A = \left(\frac{2\pi^2}{G\mu_0^2}\right)^{1/7}\left[\frac{B^4 R^{12}}{M\dot{M}^2}\right]^{1/7} \sim 80 \text{ km}. \tag{34}$$

This radius is quite large, about 7 times the NS radius (see Table 1). Here $\dot{M}$ defines the accretion rate of the free-falling overlaying material making up the NS crust, which is left out when the transition occurs (see definitions and further details in Ref.[40], and references therein). Therefore, it is quite legitimated to expect that most of the magnetic energy stored inside the magnetosphere to be drained out of the Alfvén radius. Notice in addition that the strange star radius scales as: $R_{\rm SS} = R_{\rm NS}(\frac{\rho_{\rm NS}}{\rho_{\rm SS}})^{1/3}$, where the densities ratio reads: $\frac{\rho_{\rm NS}}{\rho_{\rm SS}} \simeq 0.1 - 0.2$. Other relations between both the stars can be obtained by using conservation laws or appropriate scalings.

Table 1: Parameters used in modeling the phase transition of a neutron to a strange star, as discussed in the text.

	massa M	radius R	period P	magn. field B	accretion rate
	[$M_\odot$]	[km]	[ms]	[G]	[$M_\odot$ s^{-1}]
Neutron star	1.5	12.5	2.0	2×10^{15}	10^{-15}
Strange star	1.5	9.5-10.0	0.5	2×10^{13}	10^{-5}

Then the ESMF lines are pushed out and finally broken, in a process inverse to the standard accretion one, from r_A onwards into the SN remnant surroundings, as a violent explosion that dissipates a large part of the magnetic flux ($\Phi \sim B^2 r_A^2$) and energy trapped inside the magnetar magnetosphere[25]. Energy from the magnetic field can be dissipated via vacuum polarization and electron-positron pairs creation, as well as acceleration of charged matter flowing away (synchroton and curvature losses [41, 42]) with the explosion and material being trapped in the star magnetosphere and nearby the Alfvén radius. This is analogous to the mechanism operating during a solar flare or a coronal mass-ejection, where the very high B in a given Sun-spot is drastically diminished after flaring for a short period of time (see also Ref.[43]). In the Sun both spots outbursts and coronal mass ejections launch into space part of the Solar wind of charged and neutral particles observed to be passing by the Earth[44, 45]. In the case of an imploding NS, the phase of open magnetic field lines over which the strange star is acting as a *propeller* lasts for about $\Delta T_{prop} \sim E_{spin}/L_{prop} \sim [10^2 - 10^3]$ s, with $L_{prop} \sim 2\dot{M}c^2$ the propeller luminosity, and $E_{spin} \sim I\omega_{\rm SS}^2$ the star rotational energy. Thence, the large amount of matter ejected from the strange star at such large velocities and the pairs created, in the vacuum breakdown, drains out the dipole field of the remnant below the quantum electrodynamic limit of $B_{ce} \sim 4.4 \times 10^{13}$G [25].

To give an insight into this piece of the physics of the problem, notice that once the propeller phase is over and no more luminosity is coming from that mechanism, the magnetic field lines can recombine again if the energy released in this new stable phase is essentially the strange star rotational dipole luminosity (as the one from a millisecond pulsar), which then becomes the star dominant mechanism of energy emission, that is

$$4\frac{B_{\rm SS}^2 R_{\rm SS}^6 \omega_{\rm SS}^4}{9c^3} = I\omega_{\rm SS}\dot{\omega}_{\rm SS} \; . \tag{35}$$

For the parameters shown in Table 1 and the observed luminosity from fast rotating pulsars, this relation implies a new equilibrium magnetic field $B \sim 10^{12-13}$ G, which is below the quantum electrodynamics threshold.

Since all the differential rotation has been dragged up to build up the former ESMF, then nothing else remains to make the magnetic field to grow to its pre-collapse value. Thence no such ultra high B should reappear. We may be left with a sub-millisecond strange star[46] or a hybrid star[47] with "canonical" field strength, but no any magnetar. We note in passing that the above theoretical result is attained on the standardized assumption that the structure of the magnetic field of the pulsar is dipolar. This premise is underlying to the claim by Kouveliotou et al.[48] that a magnetar had been discovered in the soft gamma-ray repeater source SGR 1806-20. Notwithstanding, for other NS (multipolar or uniform) field configurations we do expect the overall behavior here discussed to persist, since once the spin-spin coupling is taken into account the unavoidable consequence is the appearance of a ferromagnetic (axial) configuration or structure which is dipolar in nature, and therefore the theory propound here still holds, because the magnetic field could then be amplified by a factor 10^4, putting the nascent pulsar above the threshold for stability, and the collapse ensues. Rephrasing this, one can think of this theory as a field-configuration independent constraint on initial NS magnetic field strengths.

In looking for other contexts far from those involving compact remnant stars, we noticed that recent "Tabletop Astrophysics" experiments performed by C. Wieman et al.[49] have succeeded in refining sophisticated techniques to switch atoms in a Bose-Einstein condensate (BEC) [49] from states of implosion to states of re-expansion (or explosion). In certain cases they observed that some collapses appear rather similar to microscopic supernovae. The initial implosion is followed by an explosion in which atoms are ejected as a hollow ball or in narrow jets, like in the collapse and rebound of a exploding star that forms characteristic expanding balls or streams of outflowing gas. The Wieman team named the phenomenon "Bosenovae" because of the similarities with typical supernovae. In fact, in both phenomena some material is left after the implosion as a compact object. Because of the similarity of the physics of Bose-Einstein condensates with the one we presented above, we are confident that this our theory can also be adapted to explain these impressive results by Wieman et al.[49]; Ketterle and Anglin [50]. The main feature of these oscillating Bose-Einstein condensates, said to be a scale-down version of either a neutron star or a white dwarf[49], is that for some critical fields what appears is an attractive force between the atoms and the condensate implodes and rebounds driven by some sort of internal negative pressure. We claim in this paper (a detailed description of BECs phenomenology is to be given in a work in preparation[51]) that such a negative pressure could be explained in the context of the theory introduced in this paper, since a precise relationship between gas density and magnetic field strength in the BEC is settled out by switching the atoms between attractive and non-attractive states. Thence the negative pressure acts as the equivalent of an attractive force among the atoms directed towards the magnetic field axis, leading to the BEC implosion.

11 Conclusions

We conclude by claiming that if a degenerate neutron gas is under the action of a super strong magnetic field $B_{ce} \leq B \leq B_{cn}$, for values of the density typical of NS matter its transverse pressure vanishes, the outcome being a transverse collapse. This phenomenon establishes a strong bound on the magnetic field strength expected to be found in any stable neutron star pulsar, regardless of its initial field configuration, and suggests a possible endpoint in the early evolution of highly magnetized neutron stars. They could likely be a mixed phase of nucleons and a $\pi^{\pm}, \pi^0, K^{\pm}, K^0, \bar{K}^0, \sigma, \rho^{\pm}, \omega$ meson condensate, a hybrid or strange star, or a distorted black hole but no any magnetar at all. We point out, nonetheless, that if by any mechanism a strange star could be formed directly in a supernova exlposion (which is uncertain, but not ruled out) and if vigorous dynamo action operates in the strange quark matter bulk, "pulsars" with fields higher than B_{ce} could still be formed. In other words, if any magnetars exist, they cannot be neutron stars.

12 Appendix

We use the integral representation

$$a^{1/2} = \pi^{-1/2} \int_0^{\infty} \frac{dx}{x^2}(e^{-ax^2} - 1) = \pi^{-1/2} \int_0^{\infty} dy y^{-3/2}(e^{-ay} - 1). \tag{36}$$

We regularize the divergent term dependent on a in (36), by introducing a small quantity ϵ as the lower limit in the integral, and neglecting the term independent of a

$$a^{1/2}(\epsilon) = \pi^{-1/2} \int_{\epsilon}^{\infty} dy y^{-3/2} e^{-ay}. \tag{37}$$

By taking $a(\epsilon,\eta) = p_3^2 + (\sqrt{p_\perp^2 + m_n^2} + \eta q B)^2$ and substituting in (15) and performing the Gaussian integral on p_3, one obtains

$$\Omega_{Vn}(\epsilon) = \frac{1}{4\pi^2}\sum_{\eta=\pm 1}\int_0^\infty p_\perp dp_\perp \int_\epsilon^\infty dy y^{-5/2} e^{-(\sqrt{p_\perp^2+m_n^2}+\eta qB)^2 y}. \tag{38}$$

By substituting $z = \sqrt{p_\perp^2 + m_n^2} + \eta q B$ one is left with the expression

$$\begin{aligned}\Omega_{Vn}(\epsilon) &= \frac{1}{4\pi^2}\int_\epsilon^\infty dy y^{-3} e^{-(m_n^2+q^2B^2)y}\cosh(qBmy) \\ &+ \sum_{\eta=\pm 1}\frac{\eta q B}{4\pi^2}\int_\epsilon^\infty dy y^{-2}\int_{m_n+\eta qB}^\infty e^{-z^2} y dz.\end{aligned} \tag{39}$$

By introducing the new variable $w = z - m_n - \eta q B$, the second integral in (39) becomes

$$\frac{\eta q B}{4\pi^2}\int_\epsilon^\infty dy y^{-2}\int_0^\infty e^{-[(w+m_n)^2+q^2B^2]y}\sinh(2qB(w+m_n)y). \tag{40}$$

After subtracting to $\cosh(qBmy)$ and $\sinh[2qB(w+m_n)y]$ the first two terms in their series expansion, one can take $\epsilon \to 0$ and obtain the finite expression (18). This process is equivalent to the subtraction of divergent terms, one of which is proportional to B^2, and absorbs the classical field energy term $B^2/8\pi$.

It is not difficult to check that for fields $B \ll 10^{20}$ G, the first term in (18) is the dominant one. Its first contribution after the series expansion of $\cosh(qBmy)$ is $q^3B^3m_n^3/2\pi^2(m_n^2+q^2B^2)$. For fields of order 10^{17}G such a term is of order 10^{30} ergs/cm^3, much smaller than Ω_{sn}. But for fields near 10^{20}G its contribution is comparable to that of Ω_{sn}.

Acknowledgement

The authors thanks ICTP for suppport through NET-35. The financial support of the Academy of Finland under the project No. 163394 is greatly acknowledged. HJMC thanks Fundação de Amparo á Pesquisa do Estado de Rio de Janeiro (Brasil) for the grant E-26/165.684/2003, and ICRA/CBPF for the PCI grant from the CNPq Institutional Capacitation Programme.

References

[1] M. Chaichian, S. Masood, C. Montonen, A. Pérez Martínez, H. Pérez Rojas, *Phys. Rev. Lett.* **84**, 5261 (2000).

[2] A. Pérez Martínez, H. Pérez Rojas, H. Mosquera Cuesta, hep-ph/0011399

[3] R. C. Duncan, C. Thompson, *Ap. J.* **392**, L9 (1992).

[4] V. R. Khalilov, *Phys. Rev D*, **65**, 056001-1 (2002)

[5] S. Ghosh, S. Mandal, S. Chakrabarty, astro-ph/0207492 (2002).

[6] L. Herrera and N.O. Santos, Local Anisotropy in Self-Gravitating Systems, *Phys. Rep.* **286**, 53 (1997)

[7] M.K. Mak and T. Harko, gr-qc/0110103.

[8] A. Sommerfeld, *Electrodynamics*, Academic Press, New York (1952)

[9] J.D. Jackson, *Classical Electrodynamics*, J. Wiley and Sons, New York, (1966).

[10] L. D. Landau, E. M. Lifshitz, *The Classical Theory of Fields*, Pergamon, New York, (1970)

[11] A.E. Shabad, H. Perez Rojas, *Ann. of Phys.*, **121** (1979) 432.

[12] P. Haensel, S. Bonazzola, *Astron. Astrophys.* **314**, 1017 (1996).

[13] J. E. Horvath, J. A. de Freitas Pacheco , *Int. Jour. Mod. Phys. D* **8**, 19 (1998).

[14] V.G. Bagrov, D.M. Gitman, *Exact Solutions of Relativistic Wave Equations*, Kluwer Acad. Publ. (1990).

[15] W. Heisenberg and H. Euler, *Z. Phys.* **98**, 714 (1936).

[16] L. D. Landau, E. M. Lifshitz, *Statistical Physics*, Pergamon, Oxford York, (1958)

[17] K. Huang, *Statistical Mechanics*, J. Wiley and Sons, New York, (1963)

[18] S. Weinberg, *Gravitation and Cosmology*, J. Wiley and Sons, New York, (1972)

[19] V.N. Kondratyev, *Phys. Rev. Lett.* **88**, 221101 (2002)

[20] S. Chakrabarty, D. Bandyopadhyay, S. Pal, *Phys. Rev. Lett.* **78**, 2898 (1997).

[21] H. J. Mosquera Cuesta, H. Pérez Rojas, A. Pérez Martínez, *Anisotropic pressures in very dense matter*, in preparation

[22] C. Y. Cardall, M. Prakash, J. M. Lattimer, *Astrophys. J.* **554**, 322 (2001).

[23] A. Broderick, M. Prakash, J. M. Lattimer, *Ap. J.* **537**, 351 (2000).

[24] K. Konno, T. Obata, Y. Kojima, *Astron. Astrphys.* **356**, 234 (2000)

[25] J. E. Horvath, *Int. Jour. Mod. Phys. D* **8**, 669 (1999).

[26] H. Perez Rojas, *Acta Phys. Pol.* **B17**, 861 (1986).

[27] E. S. Fradkin, Quantum Field Theory and Hydrodynamics, *Proc. of the P.N. Lebedev Inst.* No. **29**, Consultants Bureau (1967).

[28] V. Weisskopf, Kong. Dans. Videns. Selskab, *Math-fys. Meddeltser*,**14**, (1936).

[29] I-S. Suh, G. J. Mathews, *Ap. J.* **530**, 949 (2000).

[30] S. Mandal, S. Chakrabarty, astro-ph/0209462 (2002).

[31] T. Maruyama and T. Tatsumi, *Nucl. Phys. A* **693**, 710 (2001)

[32] T. Tatsumi, *Phys. Lett. B* **489**, 310 (2000)

[33] G.S. Boebinger and A. Passner, *Phys. Rev. Lett* **75** (1995) 4662.

[34] L. D. Landau, E. M. Lifshitz, *Electrodynamics of Continuous Media*, Pergamon, New York, (1963)

[35] A. Pérez Martínez, H. Pérez Rojas, H. Mosquera Cuesta, *Eur. Jour. of Phys. C*, **29**, 111-123 (2003)

[36] A. Perez Martinez, H. Perez Rojas and H. Mosquera Questa *Chin Phys Lett* (2004) Vol 21 No 11 p 2117

[37] H.-Th. Janka, E. Müller, *A. & A.* **306**, 167 (1996).

[38] I. N. Mishustin, et al., *Phys. Lett. B* **552**, 1 (2003).

[39] G. Marranghello, C. Z. Vasconzellos and J. A. de Freitas Pacheco, *Phys. Rev. D* **66**, 064027 (2002).

[40] H. J. Mosquera Cuesta, Ph. D. thesis (in portuguese), INPE, Brazil, unpublished (1998). The "Quark Nova" model can be found in R. Ouyed, J. Dey, M. Dey, *Astron. Astrophys.* **390**, L39 (2002).

[41] B. Zhang and A. K. Harding, *Astrophys. J.* **535**, L51 (2000).

[42] B. Zhang, A. K. Harding and A. G. Muslimov, *Astrophys. J.* **531**, L135 (2000).

[43] W. Kluźniak, M. Ruderman, Ap. J. Lett. 505, L113 (1998). See also M. Ruderman, L. Tao, W. Kluźniak, *Ap. J.* **542**, 243 (2001).

[44] E. N. Parker, Chin. *Journ. Astron. Astrophys.* **1**, 99 (2001).

[45] J. Chen, *Spac. Sci. Rev.* Vol.95, issue 1/2, 165 (2001).

[46] J. Madsen, *Phys. Rev. Lett.* **81**, 3311 (1998).

[47] N. K. Glendenning, *Compact stars: Nuclear physics, particle physics and general relativity*, New York, USA: Springer (1997).

[48] C. Kouveliotou, et al., Nature 393, 235 (1998), and also C. Kouveliotou, et al., *Astrophys. J.* **510**, L115 (1999).

[49] S. L. Cornish, et al., *Phys. Rev. Lett.* **85**, 1795 (2000).

[50] W. Ketterle and J. Anglin, Bose-Einstein Condensation of Atomic Gases, *Nature* **216**, 211 (2002).

[51] H. J. Mosquera Cuesta, H. Pérez Rojas, A. Pérez Martínez, *Exploding Bose-Einstein condensate and critical magnetic fields*, *Int. J. Mod. Phys. D* **14**, 1855 (2005)

In: Perspectives in Mathematical Physics
Editor: Gregory D. Monteggo
ISBN: 978-1-61122-984-4

Development of the Ignition Control Algorithm with Diagnostic Sets for an Inductive Operation in a Tokamak Reactor

Osamu Mitarai
Institute of Industrial Science and Technical Research, Kyushu Tokai University,
9-1-1 Toroku, Kumamoto 862-8652, Japan

Abstract

We review the development of a control algorithm for ignition burn for an inductive operation in a tokamak reactor including International Thermonuclear Experimental Reactor (ITER). The model based control algorithm for ignition burn is described and its control capability is demonstrated using 0-dimentional simulations. As the basic control algorithm reviewed here is based on the combination of the H-mode power threshold and the fusion power, this system has two inputs and two outputs for feedback. The external heating power is determined by the H-mode power threshold and the fueling rate is by the fusion power. For disturbances such as the change in the confinement factor including L to H and H to L transitions, helium ash particle confinement time, and impurity content, this algorithm sufficiently controls all the burn phases. The developed control algorithm, keeping the beta and the density below their limits, has a fail-safe function in the event of diagnostic failure. In addition, the D-T fuel ratio feedback and pellet injection algorithm are implemented. The diagnostic systems for ignited operation are selected and the diagnostic network system is proposed to back up the control signals for avoiding the control malfunction.

1. Introduction

Ignition analyses for the ITER tokamak have been actively conducted [1]-[7] and various diagnostic system designs have been made in parallel with those machine designs [8]-[10]. Therefore, at the present design stage of a fusion reactor including ITER, it is required to show what kind of control algorithm is to be applied and how diagnostic systems are incorporated in control systems for ignition burn. Thus, the study on integration of the ignited operation and diagnostic/control system is a challenging task for ignited operations in a fusion reactor including ITER.

Control algorithm in a fusion reactor should be as comprehensive and simple as possible for a safe operation of the machine and the nuclear licensing procedure. For this purpose, we propose the simple proportional-integration-derivative (PID) control to be the main feedback control for an ignited operation supported by various advanced control algorithms. In an ignited operation, it is preferable for the same feedback control algorithm to be applied for all the ignition burn phases such as the fusion burn rise-up, steady state and shut-down phases.

The fusion burn control algorithm should work in a non-linear plasma media, and in the thermally stable or unstable regimes with respect to the plasma density and temperature. As a usual control algorithm, such as PID control, cannot be used in the unstable regime, ignition characteristics have been studied in detail prior to creating an ignition control algorithm [11]-[13]. We have analyzed various burn conditions [12] and identified the stable path to the ignition operating point at the thermally stable regime using the preprogramming control. Thus, a design of a control algorithm becomes possible. Although the operation at the thermally unstable ignition boundary have been studied for example using neural network or nonlinear stabilization techniques [14], we do not employ the operation at the thermally unstable boundary because a fusion reactor should be operated safely and control failure must be avoided in any time.

In this paper, we review our research on the ignition control algorithm for an inductive operation in a tokamak reactor, which facilitates the diagnostic system integration design and works for the fusion power rise-up, steady state and shutdown phases with various disturbances [15]-[18]. To confirm the function of a control algorithm, zero-dimensional power balance equations have been used with the ITER93HP [19] and IPB98(y,2) [20] scalings and two H-mode power thresholds [21][22]. Based on this control algorithm, we present the ignition burn control diagram together with necessary diagnostic systems for the plasma core, where all above phases of the ignition burn can be controlled by a combination of the external heating power based on the H-mode power threshold and the fueling rate based on the fusion power.

This paper is organized as follows. In section 2, basic control algorithm for ignited operations for an inductive operation is presented. In section 3, a fail-safe control algorithm for ignited operations in an inductive operation is described. In section 4, this fuel ratio and pellet injection control algorithms are presented. In section 5, the diagnostic network system is further proposed for avoiding the control malfunction. In sections 6, further issues on control are discussed.

2. Basic Control Algorithm for Ignited Operations in Inductive Operations

The schematic basic control algorithm for ignited operations is shown in Figure1 for an inductive operation in a tokamak reactor. Here, as the plasma current is controlled by the poloidal coil power supply independently from the ignited operation, the heating power P_{EXT} and fueling S_{DT} are major two input parameters. Actually, the heating power P_{EXT} is feedback controlled using the H-mode power threshold to keep the H-mode state and the fueling rate is feedback controlled by the fusion power P_f to keep it at the set value. This is therefore the system with two inputs (heating power and fueling) and two outputs (H-mode and fusion power). In a recent tokamak system, the confinement improvement is necessary to achieve fusion output, therefore the H-mode state is the one of control target. We call this "control by a set of {P_{EXT}(HL) and $S_{DT}(P_f)$}" using two input parameters. Ignition phenomena are non-linear, and the heating power and fueling are interconnected in a fusion reactor plant as shown in Figure 1. Despite of this complicated situation, the basic control algorithm (model based feedback control algorithm) described here can control ignited operations consisting of L-H transition, various disturbances, and saturation phenomena happening during all the operation phases such as fusion burn start-up to shutdown phase. However, the operating point should be in or near the thermally stable regime with respect to the density and temperature.

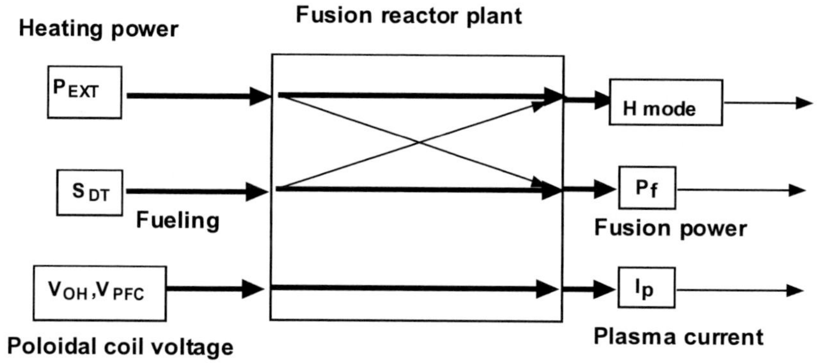

Figure 1. Schematic diagram of the basic ignition burn control algorithm with two inputs and two-output signals for an inductive operation.

2.1. Model Equations for Ignition Control

2.1.1. Control of The External Heating Power

Control of an external heating power can be made by the mathematical equation derived from the experimental data. An external heating power P_{EXT}(HL) to keep an H-mode state is given by

$$P_{EXT}(HL)\,[W] = M_{HL0}\,0.45\times10^{6}\,\bar{n}^{0.75}\,[10^{20}\,m^{-3}]\,B_t\,[T]\,R^2[m] - \left(\bar{P}_{oh}+\bar{P}_{\alpha}-\bar{P}_{b}-\bar{P}_{s}\right)V_o \tag{2-1}$$

where M_{HL0} is the set value of H-mode indicator M_{HL0}=1.05 which is larger than 1, 10^6 is the conversion factor from MW to W, $\bar{n}$ is the line averaged density, B_t is the toroidal field, R is the major radius, $\bar{P}_{oh}$ is the Ohmic heating power density, $\bar{P}$ is the alpha heating power density, $\bar{P}_b$ is the bremsstrahlung loss per unit volume, $\bar{P}_s$ is the synchrotron radiation loss per unit volume, and V_o is the plasma volume. The equation (2-1) is obtained from the H-mode power threshold given by Takizuka et al., [21]

$$M_{HL}\,(TAK) = \frac{\bar{P}_{h,net}\,[W/m^3]\,V_o[m^3]\times10^{-6}}{P_{thresh}\,(TAK)} \tag{2-2}$$

with

$$P_{thresh}(TAK) = 0.45\left\{0.6\,\bar{n}[10^{20}\,m^{-3}]R^2\right\}^{\alpha}\,\bar{n}^{-0.75}[10^{20}\,m^{-3}]\,B_t[T]\,R^2[m] \tag{2-3}$$

where $\bar{P}_{h.net}$ is the net heating power density given by $\bar{P}_{h,net}=P_{EXT}/V_o+\bar{P}_{oh}+\bar{P}_{\alpha}-\{\bar{P}_b+\bar{P}_s\}$ and 10^{-6} is the conversion factor from W to MW. In the H-mode $M_{HL}(TAK) > 1.0$ holds and in the L-mode $M_{HL}(TAK) < 1.0$. Therefore, the external heating power is applied to keep the H-mode with $M_{HL}(TAK) > 1.0$. In this study, no hysteresis and $\alpha = 0$ are assumed. Various parameters and power losses presented here are described in the references [16]-[18].

2.1.2. Control of Fueling

When a fusion power is not generated, the operation is controlled by the density. However, when fusion reaction takes place, the operation can be controlled by fueling rate which determines the fusion power through the density. This method is much safer because fueling rate is directly controlled by the fusion power output signal. Fueling rate S_{DT} is controlled by the error signal (proportional control) to maintain the total fusion power on the set value as

$$S_{DT}(P_f(t)) = S_{DT0}\,e_{DT}(P_f)\,G_{fo}(t) \tag{2-4}$$

where the error signal is

$$e_{DT}(P_f) = \left(1 - \frac{P_f(t)}{P_{fo}(t)}\right) \tag{2-5}$$

Here $P_f(t)$ is the total fusion power output signal and $P_{fo}(t)$ is the set value, the deuterium and tritium fuel ratio is assumed to be 1:1. When the fusion power signal $P_f(t)$ is smaller than the set value $P_{fo}(t)$, fueling is supplied ($S_{DT}(P_f) > 0$) because of $(1-P_f(t)/P_{fo}(t)) > 0$. When the fusion power signal is larger than the set value, the fueling is stopped ($S_{DT}(P_f) = 0$) because of $(1 - P_f(t)/P_{fo}(t)) < 0$.

In general, as integration control (I-control) can avoid the steady state offset error, and derivative control (D-control) improves the transient response, fueling rate is finally given by PID control of error signal as

$$S_{DT}(P_f) = S_{DT0}\left\{e_{DT}(P_f) + \frac{1}{T_{int}}\int_0^t e_{DT}(P_f)dt + T_d\frac{de_{DT}(P_f)}{dt}\right\}G_{fo}(t) \tag{2-6}$$

where the time constant gain of $S_{DT0} = 1\text{x}10^{19}\ m^{-3}/s$ and $G_{f0} = 50$ have been chosen throughout this paper. T_{int} is the integration time ($T_{int} = 10$ s) and T_d is the derivative time ($T_d = 0 \sim 1$ s). As the time dependent set value P_{fo} in Eq. (2-5) is given by a linearly increasing, constant and decreasing function depending on the operation phase. The fueling rate for deuterium $S_D(t)$ in the Eq. (A-1) in the Appendix is given by is $S_D(t) = 0.5\ S_{DT}(P_f)$, and for tritium in the Eq. (A-2) by $S_T(t) = 0.5\ S_{DT}(P_f)$, respectively.

When the neutral beam supplies fuel particles appreciably during heating phase, its term must be added in Eq. (2-6).

2.1.3. Control Diagram Based on a Set of {$P_{EXT}(HL)$ and $S_{DT}(P_f)$}

The basic control algorithm capable of handling all the burn phases has been found to be a control algorithm by a set of {$P_{EXT}(P_f)$ and $S_{DT}(P_f)$} [15][16]. Diagnostic systems for ignited operations should be designed to estimate these quantities. To estimate the heating power P_{EXT} and the fueling rate S_{DT}, the following plasma parameters appeared in Eqs. (2-1) and (2-6) should be measured, such as the line averaged density $\overline{n}$, the toroidal field B_t, the Ohmic heating power $\overline{P}_{oh}$, the alpha heating power $\overline{P}_{□}$, the bremsstrahlung loss $\overline{P}_b$, and the synchrotron radiation loss $\overline{P}_s$.

In Figure 2 is shown the schematic diagram of the feedback control and diagnostic systems for the core burn control in a tokamak reactor. This control algorithm is consisting of two inputs and multi-output signals. Two inputs are the external heating power P_{EXT} and the fueling rate S_{DT}. Multi-output signals are to obtain two output signals of {M_{HL}, P_f} for feedback control. The fusion power P_f is obtained by $P_f = P_n + P_\alpha = 1.2\ P_n$ which is measured by the neutron diagnostics [23][24] and alpha heating power estimated by $P_\alpha = P_n/5$. Output H-mode indicator M_{HL} is estimated using the actual heating power $P_{NET} = \overline{P}_{h,net}V_o = P_{EXT} + [\overline{P}_{oh} + \overline{P}_\alpha - \{\overline{P}_b + \overline{P}_s\}]V_o$ and the threshold power P_{thresh} by Eq. (2-2). If the lost alpha fraction $(1 - \eta_\alpha)$ could be measured, the actual alpha heating power $P_\alpha(\eta_\alpha)$ is estimated. Bolometer array measures the total radiation loss power P_B including the bremsstrahlung and line radiations [25] and electron cyclotron emission spectrum should be measured for the synchrotron radiation loss P_S. Here, the Ohmic heating power P_{OH} is estimated by the plasma current I_p and the loop voltage V_L.

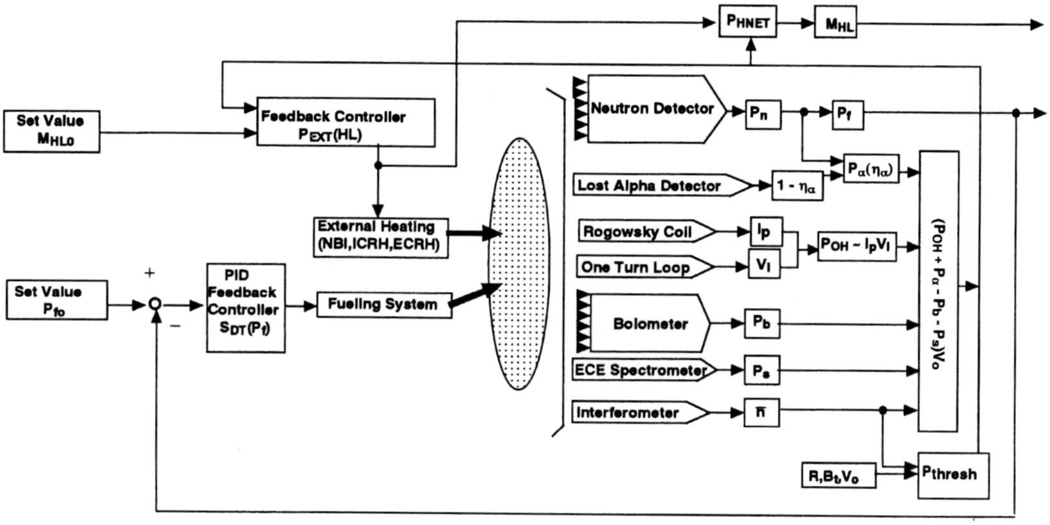

Figure 2. Basic ignition burn control algorithm with two inputs {P_{EXT}, S_{DT}}and multi-output signals for two outputs {M_{HL}, P_f} for feedback control.

The time response of the diagnostic and actuator systems required for this ignition burn control is ~ 100 ms, but 10 ms has been used for calculation in this study. This has been confirmed for various ignition burn scenarios without sawtooth and abnormal events.

Although the shorter time response of each diagnostics may be necessary to measure plasma parameters for physics understanding and disruption related phenomena [9], 100 ms or less time resolution is a rough measure for the global ignition control.

2.2. Simulations Supporting the Basic Control Algorithm

Simulations to justify the control algorithm have been conducted using the ITER-class tokamak reactor as listed in Table 1. Two cases are different with respect to the impurity content, the effective alpha ash confinement time to the energy confinement time, the confinement time scaling and the H-mode power threshold.

Table 1. Assumed machine and ITER-class plasma parameters in this analysis.

Major radius	R= 8.14 m	
Minor radius	a = 2.8 m	
Toroidal field	B_t = 5.68 T	
Elongation	κ = 1.6	
Plasma current	I_p = 21.0 MA	
Impurity fraction	n_{Be}/n_e= 2 %	n_{Be}/n_e= 2 %, n_{Ar}/n_e= 0.2 %
Alpha particle confinement time	τ_α^*/τ_E= 10	τ_α^*/τ_E= 5
Temperature ratio	T_i/T_e = 0.95	
Fuel ratio	n_T/n_D = 1	
Alpha particle heating efficiency	η_α = 0.97	
Wall reflectivity for synchrotron radiation loss	R_{eff} = 0.9	
Hole fraction for synchrotron radiation loss	f_H = 0.1	
Density profile	α_n = 0.15	
Temperature profile	α_T= 0.9	
Enhancement factor	γ_{HH} =1.0 (ITER93HP)	γ_{HH} =1.0 (IPB98(y,2))
Peak electron density	n(0) ~ 1.15x10^{20} m^{-3}	n(0) ~ 1.23x10^{20} m^{-3}
Peak ion temperature	$T_i(0)$ ~ 25 keV	$T_i(0)$ ~ 18.5 keV
Alpha ash fraction	$f_{□}$ ~ 14.8 %	$f_{□}$ ~ 5.7 %
Effective ion charge	Z_{eff} ~ 1.54	Z_{eff} ~ 1.95
H-mode power threshold	Takizuka et al [21]	Righi et al [22]

2.2.1. Control of Ignition Access

2.2.1.1. Ignition Access with γ_{HH} = 1.0

The plasma current is inductively established by the Ohmic transformer, and the fusion power is ramped-up by the heating power and fueling. Initially the plasma has a L-mode state, and the L-H transition takes place when the net heating power P_{NET} exceeds the threshold power P_{thresh}. This control algorithm does work in these situations. In Figure 3 is shown the temporal evolution of plasma parameters during the ignition access phase. The preprogrammed external heating power of 60 MW is initially applied to the Ohmic plasma for ~ 4 sec to induce the L-H transition. When it exceeds the threshold power P_{thresh} (dash-dot line), the L-H transition takes place and the feedback control of the heating power is turned on. After transition to the H-mode as identified by $M_{HL} > 1$, the plasma conduction loss is decreased due to the confinement improvement from $\gamma_{HH} = 1/3$ to 1.0, where the net heating power P_{NET} is still larger than the threshold power P_{thresh}. After the H-transition, the net heating power is always larger than the threshold power and reaches the steady state value. The confinement time is gradually decreased during the L-mode phase, and then suddenly increased due to L-H transition.

The heating power of ~ 60 MW is gradually reduced to zero to keep the operating point at M_{HL} ~ 1.05. After switching off the heating power, the H-mode indicator M_{HL} increases with fusion power. It is seen that the fusion power P_f is almost identical with the set waveform of the fusion power P_{fo} as shown by the dotted line and the fusion power rise time can be controlled precisely. The steady state ignited operating point thus can be reached by this control algorithm after a long time depending on the alpha ash build-up, where parameters are $n(0) \sim 1.15 \times 10^{20}$ m^{-3}, $T(0) \sim 25.0$ keV with $M_{HL} = 1.05$. The fusion power rise-up time can be precisely controlled using the feedback control algorithm proposed here. Even longer fusion power rise-up time is also possible by this control method.

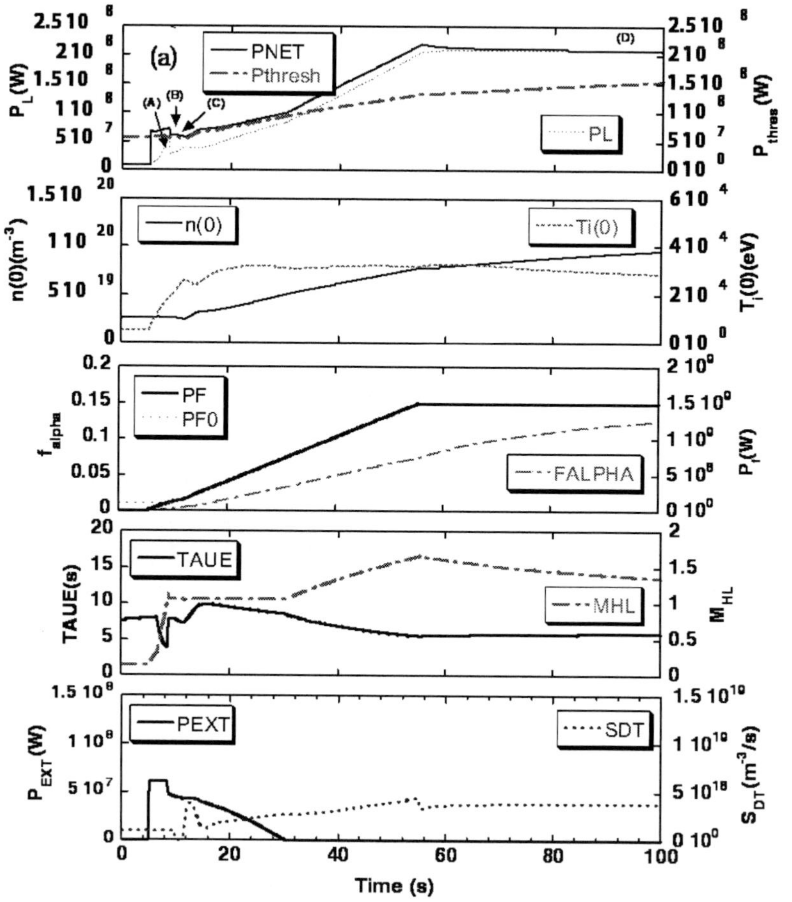

Figure 3 (Continued).

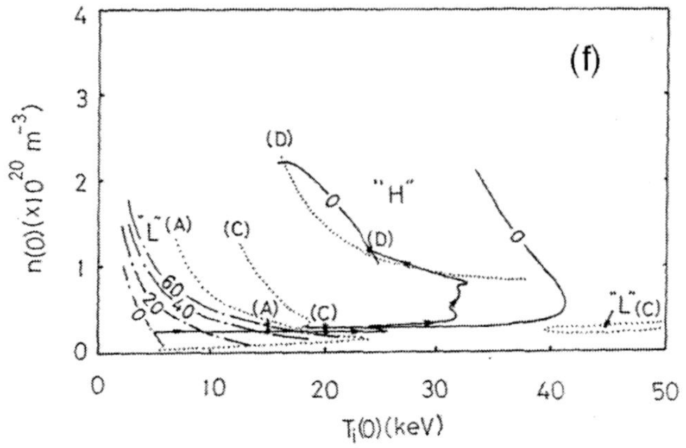

Figure 3. Temporal evolution of plasma parameters during the ignition access phase. (a) P_{NET} the net heating power, P_{thresh} the H-mode threshold power, P_L the plasma conduction loss, (b) $T_i(0)$ the peak ion temperature and n(0) the peak electron density, (c) P_F the fusion power and f_{alpha} is the helium ash fraction, (d) τ_E confinement time and M_{HL} the H-mode indicator, and (e) P_{EXT} the external heating power and S_{DT} the total fueling rate. No hysteresis effect is assumed after H-transition.

The initial fusion power was assumed to be $P_{f0} = 0.1$ GW as indicated by the dotted line to remove the singularity in the fueling term of $S_{DT0}G_{fo}(1 - P_f/P_{f0})$. The feedback control of the fueling was switched on at ~ 10 sec when the fusion power reaches the initial target value. The initial rise-up phase is sensitive to the switching-on time of the feedback control and the value of gain G_{fo}. The time dependent gain or bump-less control may improve this situation.

(f) POPCON and the ignition access path (solid line with (A),(C) and (D)). The dotted lines with (A), (C) and (D) show the L-H boundaries corresponding to them in (a). The solid line with "—0—" is the ignition boundary with P_{EXT}=0 MW.

The confinement time scaling is the ITER93HP scaling [19] with γ_{HH}=1.0 and the H-mode power threshold given by by Takizuka et al. [21] with M_{HL0}=1.05. (Figure5 in the reference [26]).

As described before, the ignition burn control does work in the thermally stable regime. When ignition access path is drawn in POPCON as shown in Figure 3-(f), it is seen that the operating point proceeds to the high temperature regime at first and then to the lower temperature with the higher density, and reaches the thermally stable ignition boundary as shown by the points "(A)" → "(C)" → "(D)" on the trace. Ignition control algorithm is thus designed by the set of {$P_{EXT}(P_f)$ and $S_{DT}(P_f)$} a.

2.2.1.2. Sub-ignition Access with γ_{HH} < 1.0

When the confinement factor is smaller, it is difficult to achieve ignition, but this control algorithm works properly. Actually for the smaller confinement factor γ_{HH} <1.0, the external heating power becomes finite and sub-ignited condition is maintained at M_{HL} = 1.05. Although oscillation sometimes appears in the plasma parameters, it could be removed using the maximal value of fueling (saturation effect), which is discussed in detail in the reference [16].

2.2.2. Response to Disturbance in the Steady State Phase

2.2.2.1. Fusion Power Surge (Over-Ignition) by Sudden Increase in the Confinement Factor

When disturbance such as the confinement improvement somehow takes place during ignited operation, this basic control algorithm can handle it. However, as ignition regime is inherently unstable, it is not possible to control the operating point completely.

For example, when the confinement factor suddenly increases from γ_{HH} = 1.0 to 2.0 at the ignition operating point, the fusion power increases from 1.45 GW to ~ 1.9 GW as shown in Figure 4. Fueling is stopped to reduce the fusion power, but the fusion power surge is not completely suppressed due to thermally unstable nature in the ignition regime. The confinement time increases from 6.45 sec to ~ 12.5 sec as set by the confinement factor, and the alpha ash density fraction increases due to increases in the alpha ash confinement time τ_α^*. The H-mode eventually returns to the L-mode after 16 sec due to fuel dilution by alpha ash. As it is assumed that the confinement factors are set at γ_{HH} = 2.0 for the H-mode and 1/3 for the L-mode during this phase, cyclic H-L-H transitions take place with the period of ~ 0.9 sec. Thus, the operating point is seen to stay near the H-L boundary after the first H-L transition. The plasma conduction loss P_L is decreased initially because the confinement time is increased, but it increases up to ~ 820 MW at the later H-L transition of 317.5 sec, which is excessively large. Thus we see that this control algorithm works sufficiently in the thermally unstable ignition regime and the L-H transition regime. To suppress the fusion power surge completely, impurity is to be injected additionally.

In the case of various disturbances such as changes in impurity fraction, particle confinement, energy confinement etc, this basic control algorithm can be used. Also without knowledge of alpha ash fraction, ignition can be controlled by this control algorithm.

The confinement time scaling is the ITER93HP scaling [19] with γ_{HH}=1.0 and the H-mode power threshold given by by Takizuka et al. [21] with M_{HL0}=1.05. (Figure 7 in the reference [26])

2.2.2.2. Sub-ignited Operation by Sudden Decrease in Confinement

When the confinement factor decreases suddenly, the heating power is applied to keep the H-mode and fusion power, and then ignition becomes sub-ignition. In this case this basic control algorithm works properly, however, the plasma parameters such as fusion power etc sometimes oscillate. To suppress these oscillations the additional modified control algorithm was examined[16].

2.2.3. Control of Shutdown Phase

In general, it is desirable to decreases the fusion power linearly without sudden parameter change. Without the H-mode operation, it is difficult to decrease the fusion power gradually because the fusion power quickly drops in the L-mode. Therefore, this basic control algorithm by a set of {P_{EXT}(HL) and $S_{DT}(P_f)$} with the heating power to maintain the H-mode works properly during the shutdown phase of the fusion power.

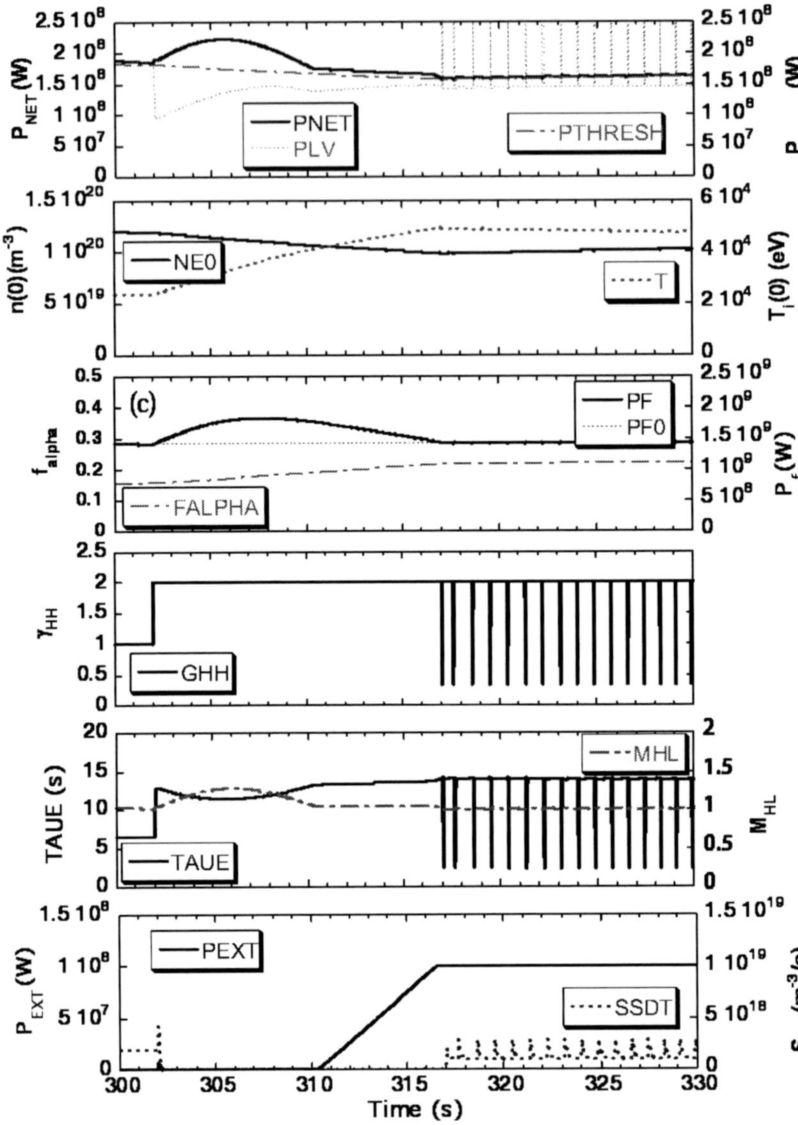

Figure 4. The effect of the sudden change in the confinement factor from $\gamma_{HH} = 1$ to 2 at 302 sec on the temporal evolution of plasma parameters. (a) P_{NET} the net heating power, P_{thresh} the H-mode threshold power, P_L the plasma conduction loss, (b) $T_i(0)$ the peak ion temperature and n(0) the peak electron density, (c) P_f the fusion power and f_{alpha} is the helium ash fraction, (d) γ_{HH} the confinement factor, (e) τ_E confinement time and M_{HL} the H-mode indicator, and (f) P_{EXT} the external heating power and S_{DT} the total fueling rate.

In Figure 5 is shown the fusion burn shutdown phase. The shutdown time interval was set to 100 sec. In this control algorithm the heating power is applied to maintain the H-mode during the shutdown phase, and then increases to the maximum value of 100 MW. The H-mode indicator M_{HL} is around unity after 1070 sec and P_{NET} is close to the threshold power P_{thresh} during 100 MW heating phase. Thus, H-L-H cyclic transitions take place and the confinement time swings between H- and L-modes with the confinement factor of $\gamma_{HH} = 1$

and 1/3, respectively. At this time, the plasma conduction loss P_L repeatedly increases up to ~ 380 MW due to a sudden decrease in the confinement time. The overall confinement time gradually changes to the Neo-Alcator confinement time after decay of the fusion power. On the other hand, the net heating power slowly decreases proportional to the fusion power. Care must be taken during the H-L-H transition phase, because the heat load to the divertor plate is increased by a factor of two over the steady state value.

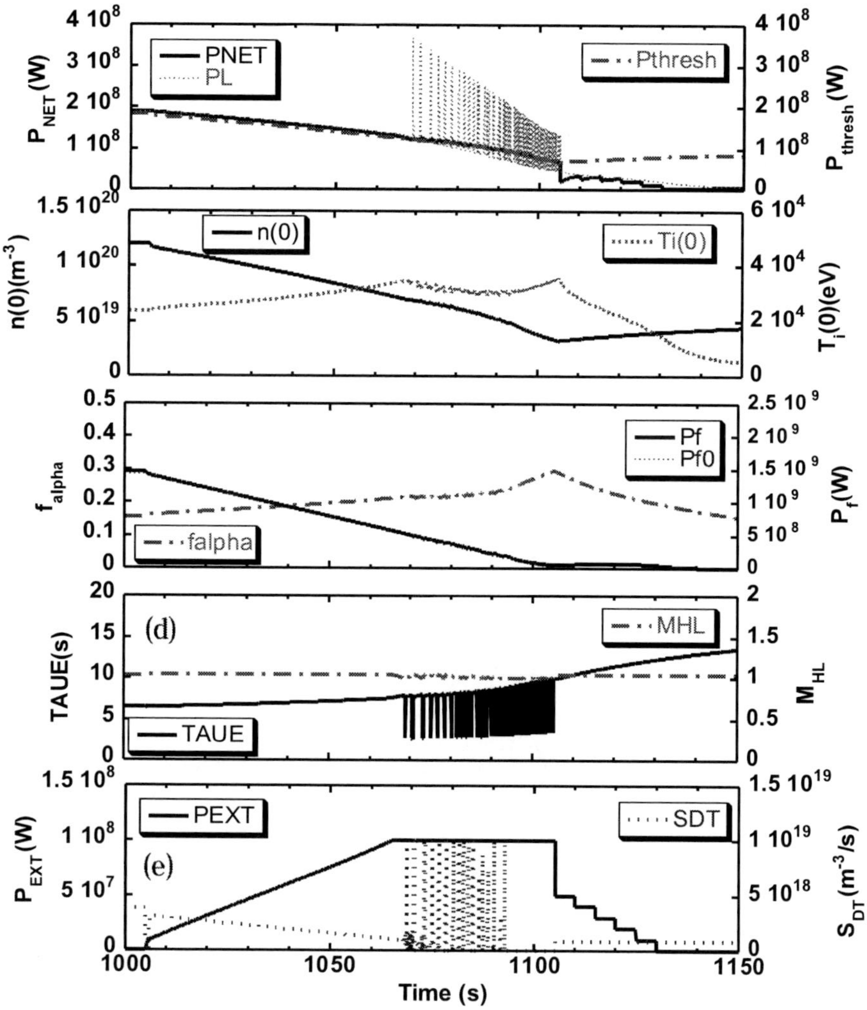

Figure 5. Temporal evolution of plasma parameters during the fusion burn shutdown phase with a shutdown time of 100 sec. H-L-H transitions are observed around M_{HL}~ 1. Each figure box is the same as in Figure 3. The confinement time scaling is the ITER93HP scaling [19] with γ_{HH}=1.0 and the H-mode power threshold given by by Takizuka et al. [21] with M_{HL0}=1.05.(Figure 8 in the reference [26]).

2.3. Variation of Control Algorithm

Variations of control algorithms as given by a set of {max{P_{EXT}(HL), $P_{EXT}(P_f)$} and $S_{DT}(P_f)$} or { $P_{EXT}(P_f)$} and $S_{DT}(P_f)$} are also possible for limited situations. The first one can be used to improve the basic control algorithm for the case of the sudden change in the confinement factor to lower one. As the fusion power alone cannot control the ignition burn rise-up phase, the second one may be possible to use during the steady state phase and shut down phases as demonstrated previously [16]. However, this second algorithm is not so good as the first one.

3. Fail-Safe Control Algorithm for Ignited Operations in Inductive Operations

3.1. Operation within the Performance Limitation by {P_{EXT}(HL) and min{$S_{DT}(P_f)$, $S_{DT}(\bar{n})$, $S_{DT}(\langle\beta\rangle)$}}

Plasma performance is degraded when the beta or density exceeds their limits. As the basic control algorithm described in Figure 2 cannot handle these cases, we further developed the control algorithm working within the beta or the density limit [16].

While the heating power is controlled by the H-mode power threshold, fueling is now controlled by the one of fusion power, density and beta value. Actually, the minimum value in three parameters determines the fueling rate. Thus, the control algorithm is described by a set of {P_{EXT}(HL) and min{$S_{DT}(P_f)$, $S_{DT}(\bar{n})$, $S_{DT}(\langle\beta\rangle)$}} where

$$S_{DT}(\bar{n}) = S_{DT0}\, e_{DT}(\bar{n})\, G_{fo} \tag{3-1}$$

$$S_{DT}(\langle\beta\rangle) = S_{DT0}\, e_{DT}(\langle\beta\rangle)\, G_{fo} \tag{3-2}$$

where

$$e_{DT}(\bar{n}) = \left(1 - \frac{\bar{n}}{\bar{n}_{GW}}\right) \tag{3-3}$$

and

$$e_{DT}(\langle\beta\rangle) = \left(1 - \frac{\langle\beta\rangle}{\langle\beta\rangle_{max}}\right) \tag{3-4}$$

Here $\bar{n}_{GW}$ is the line averaged density limit such as the Greenwalt limit [27] and $\langle\beta\rangle_{max}$ is the volume averaged maximum beta value. As the error signals based on the averaged value, it can be used when the profile is changed. After selecting the minimum error signal by

$$e_{DT}(\min) = \min\left\{ e_{DT}(P_f),\ e_{DT}(\bar{n}),\ e_{DT}(\langle\beta\rangle) \right\} \tag{3-5}$$

the total fueling rate for PID control is given by

$$S_{DT}(\min) = S_{DT0}\left\{ e_{DT}(\min) + \frac{1}{T_{int}}\int_0^t e_{DT}(\min)dt + T_d\frac{de_{DT}(\min)}{dt} \right\} G_{fo}(t) \tag{3-6}$$

The control diagram for this case is shown in Figure 6). As the fueling rate is controlled by the one of three parameters P_f, $\bar{n}$ and <□>, the neural network to predict the density limit and beta limit would be additionally implemented in the control diagram [18]. This control algorithm is also robust against the failure in these diagnostics such as neutron detectors, interferometer and diamagnetic loop. This is because even if one of three diagnostics has a failure, the other diagnostics regulate the fusion power through fueling [17]. In this sense, we call this the fail-safe control algorithm.

This fail-safe control algorithm is very convenient because it can be used in the H/D operation without fusion burn, which would be conducted prior to D-T operation. In this time, the fusion power P_f should be set at the very large value not to affect the minimum value of fueling $\{S_{DT}(P_f), S_{DT}(\bar{n}), S_{DT}(\langle\beta\rangle)\}$. The fueling is now controlled by either the density or beta. Or the fusion power feedback signal line should be completely removed from the control diagram to obtain $\min\{S_{DT}(\bar{n}), S_{DT}(\langle\beta\rangle)\}$.

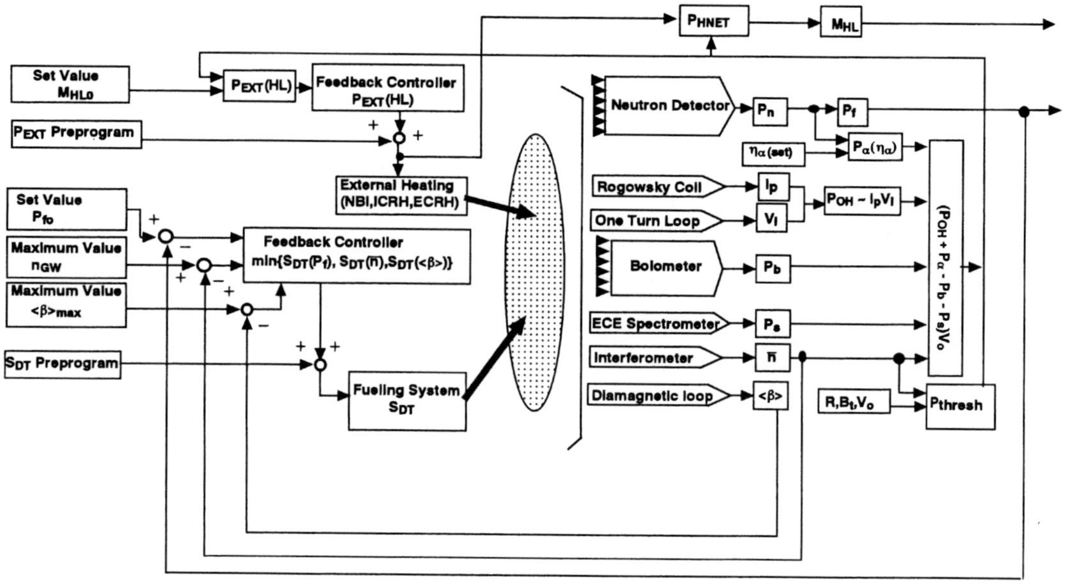

Figure 6. Ignition burn control algorithm working below the performance limitation on the density and beta by a set of $\{P_{EXT}(HL)$ and $\min\{S_{DT}(P_f), S_{DT}(\bar{n}), S_{DT}(\langle\beta\rangle)\}\}$.

When the density limit n_{GW} is set at the relatively lower value of $1.05x10^{20}$ m^{-3} and the maximum beta value is set at the large value of $<\beta>_{max} = 8.0$ % in the case of the IPB98(y,2) confinement time scaling [20] ωιτη $\gamma_{HH} = 1.0$, then the fusion power is regulated at $1.05x10^{20}$ m^{-3} by the density feedback control as shown in Figure 7. Therefore, the fusion power is ~ 1.3 GW less than 1.5 GW.

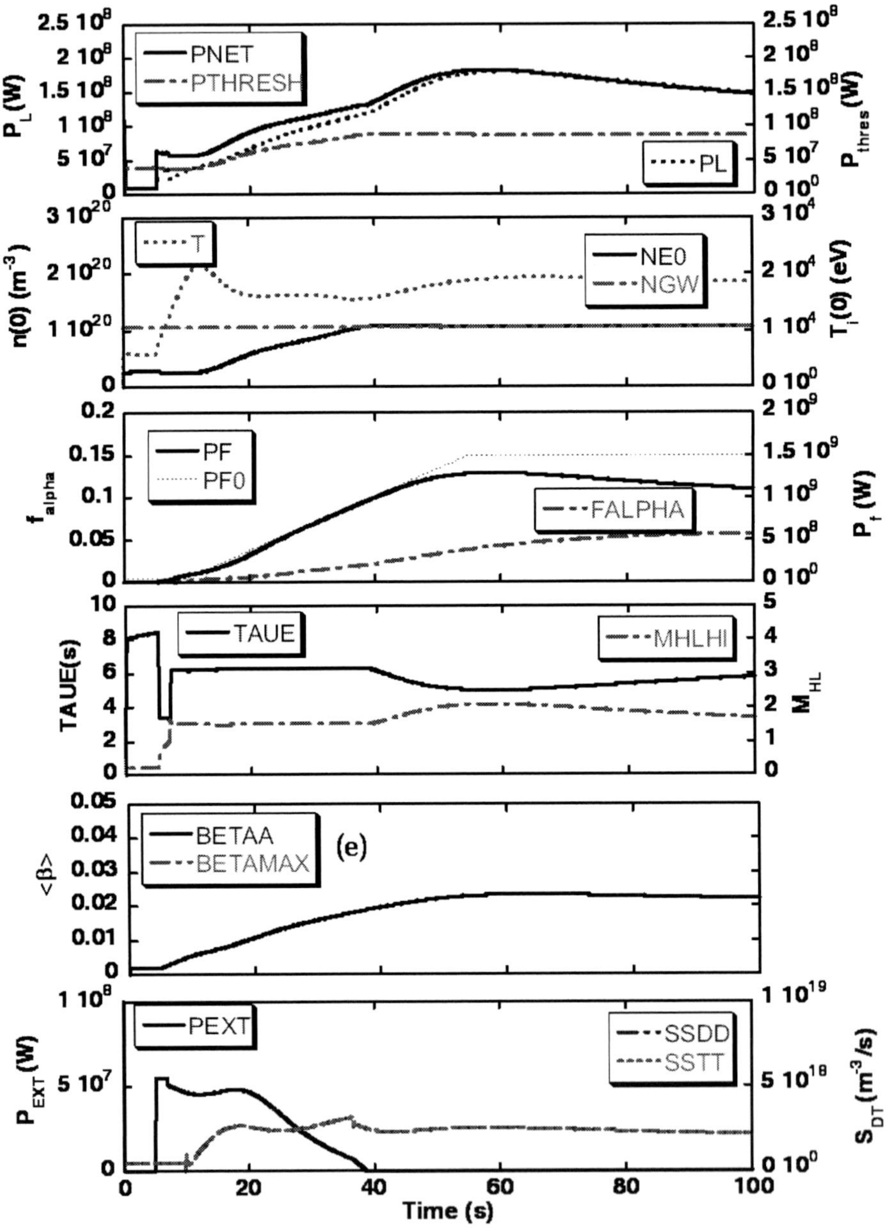

Figure 7. Ignited operation regulated by the density limit of $n_{GW}=1.05x10^{20}$ m^{-3} as seen in (b) for $<\beta>_{max}$ = 8.0 %. The confinement time scaling is IPB98(y,2) scaling with $\gamma_{HH}=1.0$ and the H-mode power threshold given by Righi et al [22] with $M_{HL0}=1.5$.

On the other hand, when the density limit n_{GW} is set at the higher value of 2.1×10^{20} m^{-3} and the maximum beta value is set at the lower value $<\beta>_{max} = 2.3$ %, the beta value is almost kept constant by the beta limit feedback control, and then the smaller fusion power of ~1.3 GW is obtained as shown in Figure 8.

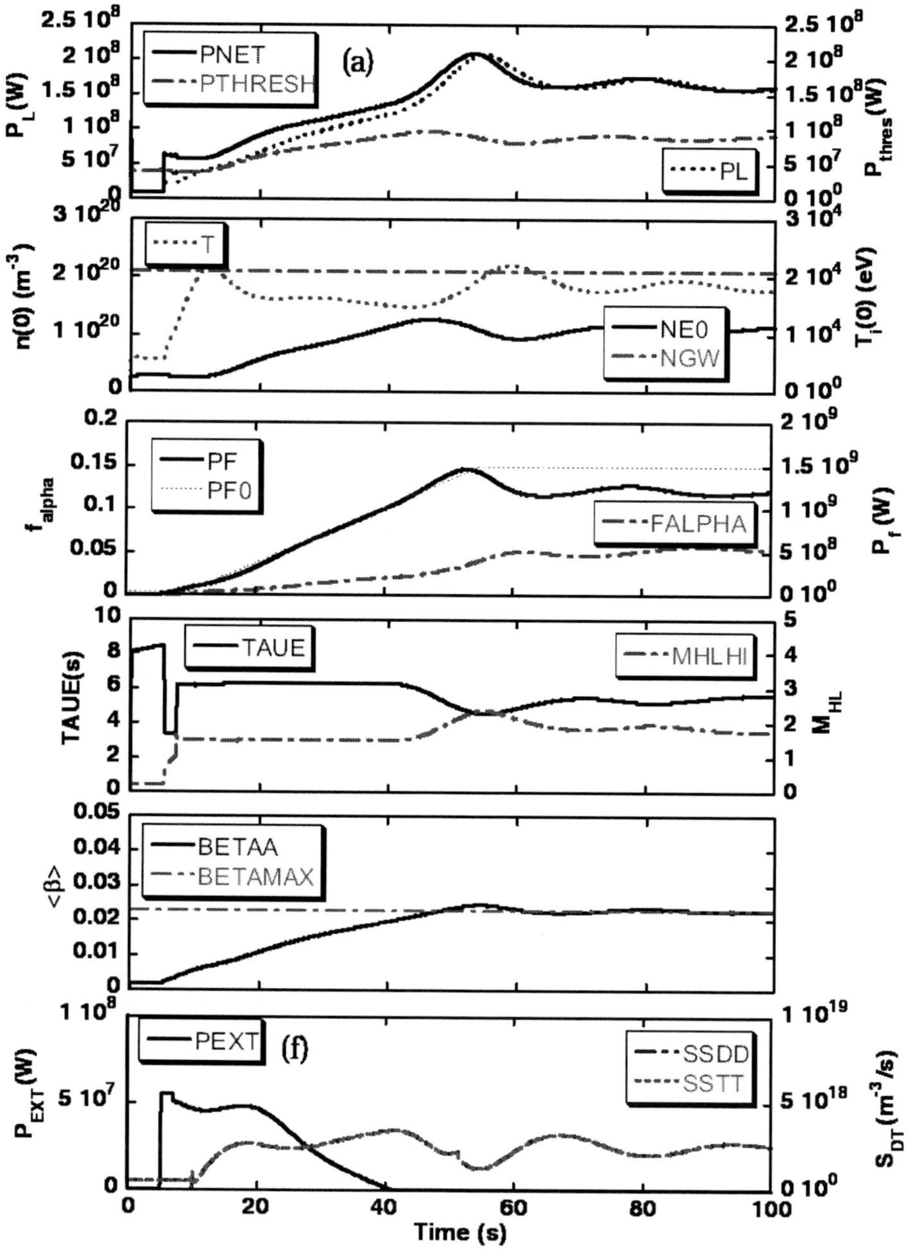

Figure 8. Ignited operation regulated by the beta limit of $<\beta>_{max}=2.3$ % as seen in (e) for $n_{GW}=2.1 \times 10^{20}$ m^{-3}. The other box is the same as in Fig, 3. The confinement time scaling is IPB98(y,2) scaling with $\gamma_{HH}=1.0$ and the H-mode power threshold given by Righi et al [22].

3.2. Measurement Errors and Fail-Safe Function in the Control Algorithm

Replacing the fusion power P_f, the line average density $\bar{n}$ and the toroidal beta <β> signals in Eqs.(2-5), (3-3) and (3-4) by $P_f(1+\varepsilon_{pf})$, $\bar{n}\,(1+\varepsilon_n)$ and $<\beta>(1+\varepsilon_n)$ with ε_{pf}, ε_n, ε_β being the each error, respectively, the effect of measurement errors, on ignited operation can be discussed[17] together with its fail-safe function.

Ignited operation must be made safely. From this point of view we have studied what happens when these signals are lost during ignited operation. A loss of signal event (LOSE) in these signals P_f, $\bar{n}_{sig}$ and <β> would produce severe problems in the ignited operation. For example, if a complete loss of signal event (LOSE) (ε_{pn} = - 1.0) in neutron diagnostics happens in the fusion burn phase, a fusion power immediately increases by sudden increase in the fueling and the heating power because the feedback signal of fueling becomes maximum as $S_{DT} = S_{DT0}G_{fo}$ due to the null signal $P_f(1 + \varepsilon_{pn}) \rightarrow 0$ in the proportional feedback control in Eq. (2-5), and the heating power P_{EXT}(HL) in Eq. (2-1) is increased to the maximum value of 100 MW due to $P_\alpha\,(1 + \varepsilon_{pn}) \rightarrow 0$. In these cases, the fail-safe function in the control algorithm can suppress fusion power surge by the other signals of $\bar{n}$ or <β>. Various cases have been examined in the reference [17], and this control algorithm shows the capability of the safe ignited operation in a reactor.

4. Fuel Ratio Control and Pellet Injection Algorithm for Ignited Operations

4.1. Control of D-T Fuel Ratio

When deuterium and tritium fueling rate are unbalanced, or when the deuterium and tritium particle confinement times including the recycling at the first wall and the pumping effect are different, the deuterium and tritium density would be different, and the electron density increases and exceeds the density limit for the same fusion power. To avoid fuel ion unbalance, the fuel ratio feedback control is to be implemented. On the other hand, the fuel ratio measurement diagnostics are not yet developed, it is assumed here that such diagnostics works properly.

The fuel ratio control is connected to the total fueling S_{DT}(min) in cascade. The equations for the proportional-integral (PI) control of the fuel ratio are given as follows. Using the error signals of the fuel ratio for the set value of the fuel ratio n_D:n_T=1:1 or $n_D/(n_D+n_T)$=1/2,

$$e_D(n_D/n_T) = \left(\frac{1}{2} - \frac{n_D}{n_D + n_T} \right) \tag{4-1}$$

$$e_T(n_D/n_T) = \left(\frac{1}{2} - \frac{n_T}{n_D + n_T} \right) \tag{4-2}$$

deuterium and tritium fueling rates are given by

$$S_D(n_D/n_T,\ \min) = \left\{ \frac{1}{2} + G_{NDT}\left(e_D(n_D/n_T) + \frac{1}{T_{NDTint}} \int_0^t e_D(n_D/n_T)dt \right) \right\} S_{DT}(\min)$$

(4-3)

$$S_T(n_D/n_T,\ \min) = \left\{ \frac{1}{2} + G_{NDT}\left(e_T(n_D/n_T) + \frac{1}{T_{NDTint}} \int_0^t e_T(n_D/n_T)dt \right) \right\} S_{DT}(\min)$$

(4-4)

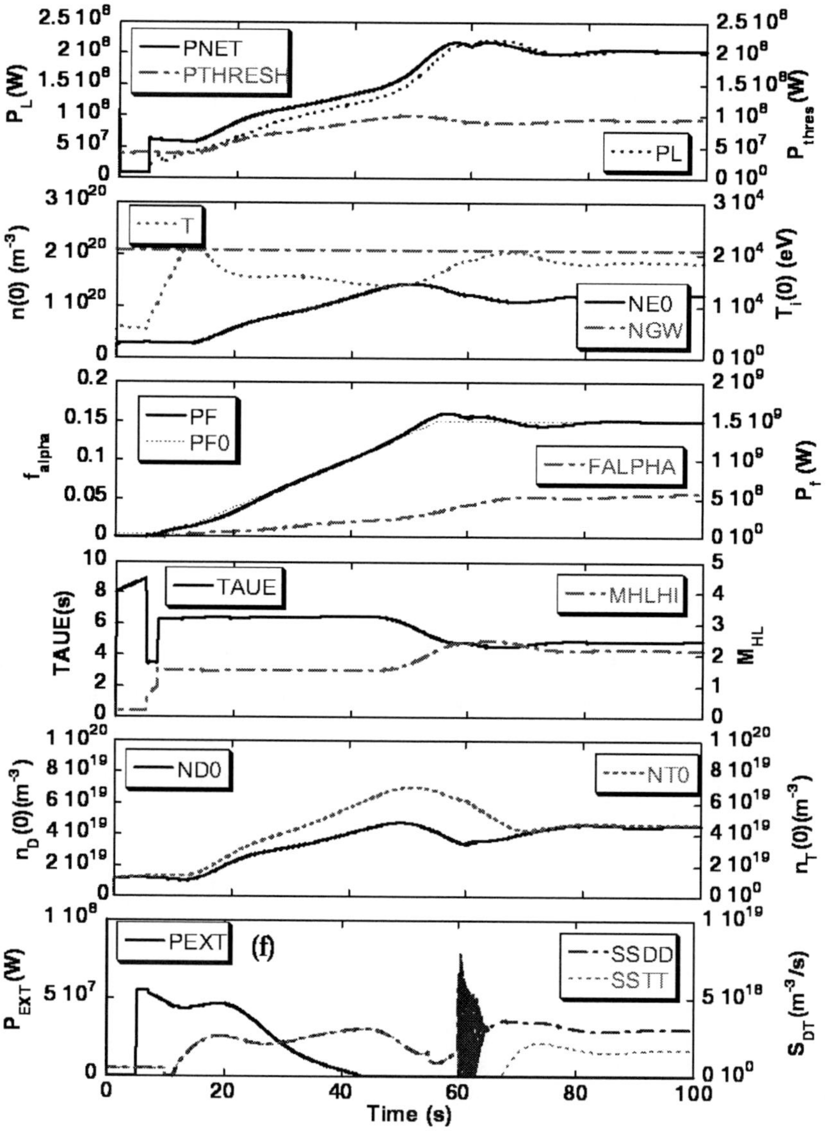

Figure 9. Ignited operation in the case of the unbalanced fuel particle confinement time of $\tau^*_T/\tau^*_D = 2$ for n_{GW}=2.1x10^{20} m^{-3} and $<\beta>_{max}$ = 8.0 %. Deuterium and tritium fuel ion densities are diverted initially.

as seen in (e). As the fuel ratio feedback control was switched on at 60 s, the fuel ion densities become equal after 10 s. The deuterium ("SSDD") and tritium ("SSTT") fueling rates are different after 60 s as seen in (f). The confinement time scaling is IPB98(y,2) scaling with γ_{HH}=1.0 and the H-mode power threshold given by Righi et al [22].

We have used only PI control to demonstrate its performance. We have used G_{NDT} = 10 and T_{NDTint} = 100 sec in this study.

Simulations using above fuel ratio feedback control algorithm is shown in Figure 9 for the unbalanced particle confinement time of the tritium and deuterium as $\tau^*_T/\tau^*_D = 2$. Before 60 sec the deuterium "ND0" and tritium "NT0" ion densities are diverted due to the difference of the particle confinement times as seen in Figure 9-(e). After the fuel ratio feedback control was switched on at 60 sec, both fuel ion densities become equal in 10 s by the fuel ratio feedback control.

This fuel ratio control algorithm is additionally implemented in cascade after the total fueling control of min{$S_{DT}(P_f)$, $S_{DT}(\overline{n})$, $S_{DT}(<\beta>)$} in the fail-safe control diagram as shown in Figure 10. The fuel ratio measurement system (n_T/n_D measurement) is providing the information to the feedback system.

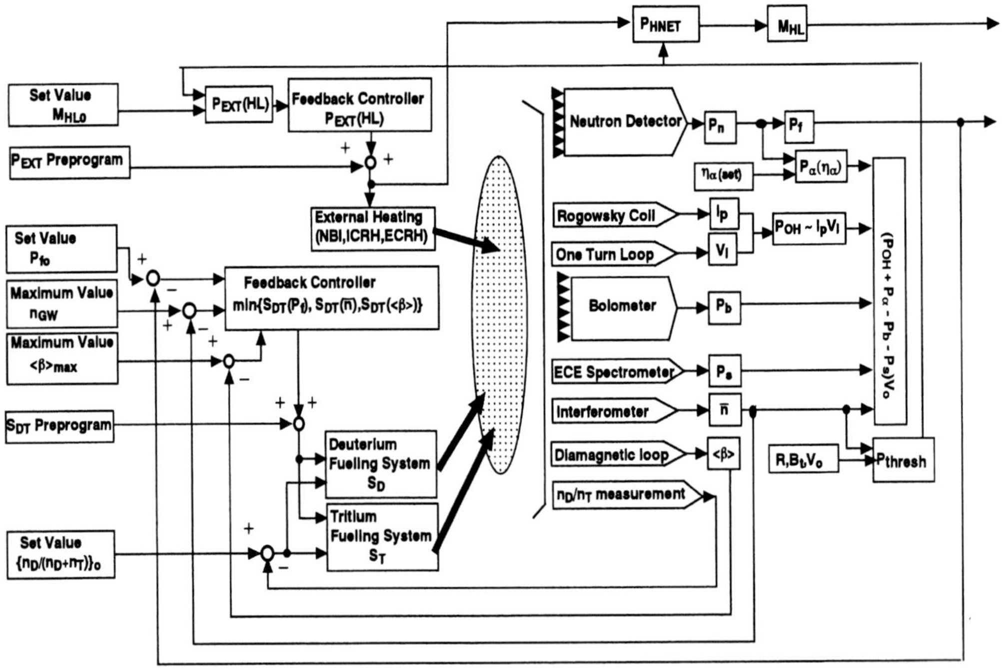

Figure 10. Fail-safe ignition control algorithm with D-T fuel ratio control. Three inputs {P_{EXT}, S_D, and S_T} and two outputs {M_{HL}, P_f} system.

However, if the fuel ratio diagnostics is not developed in time, the alternative control method should be considered for obtaining the optimum fuel ratio. When the fuel ratio unbalance is increased, the required electron density increases for the fusion power of 1.5 GW as shown in Figure 11. At the optimum fuel ratio of $n_D/(n_D+n_T)$=0.5, the density is the minimum. Therefore, when deuterium is injected, and if the density decreases for the same fusion power, it is going to fuel balancing. If the density increases for the same fusion power, it is going to fuel unbalancing.

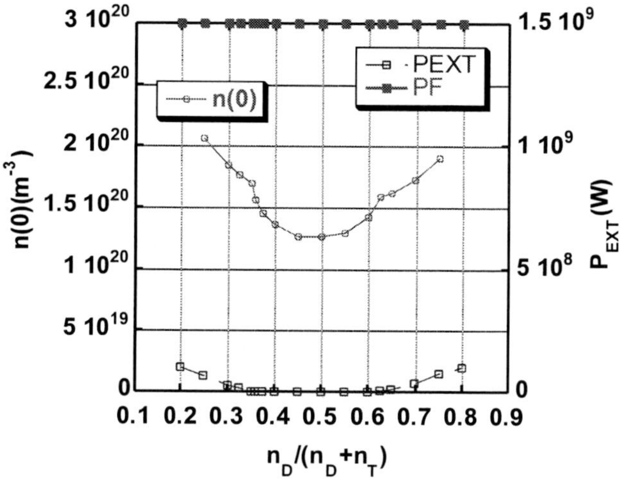

Figure 11. The fuel ratio vs. the required density for 1.5 GW fusion power. Even without fuel ratio diagnostic system, ignited operation may be possible for the fuel ratio of $n_D/(n_D + n_T) = 0.35 \sim 0.65$ if the high density operation is allowed. The confinement time scaling is IPB98(y,2) scaling with γ_{HH}=1.0 and the H-mode power threshold given by Righi et al [22] with M_{HL0}=1.

4.2. Pellet Injection Algorithm

To achieve the high density, ice pellet of the deuterium and tritium is injected repetitively. In general, the fusion power in a reactor oscillates by repetitive pellet injections. Therefore, this oscillation must be reduced as small as possible. For this purpose the high frequency pellet injection is used together with the slow gas puffing. This effectiveness has been also confirmed in the FFHR helical reactor [28].

The centrifugal pellet injector is suitable for pellet injection because of inherent cyclic operation. The centrifugal pellet injector with 1 to 15 Hz pellet injection capability has already demonstrated experimentally [29]. The pellet injection algorithm is added to D-T gas puffing algorithm as shown in Figure 12 [28]. D-T pellet chopping unit, which cuts the frozen DT pellet with the 1:1 fuel ratio, could be controlled by the electric signal generated by the fusion power error signal $e_{DT}(P_f)$ (Eq.(2-2)). When the fusion power oscillates, slow gas puffing controlled by the same fusion power can compensate the slow variation together with the higher frequency injection f_{INJ} as will be shown in Figure 13.

Example of the ignited operation by the pellet injection is shown in Figure 13. Here the pellet is assumed to penetrate into the plasma core and the fueling rate is corresponding to the pellet size. The fuelling rate by pellet is assumed as the square wave with $S_{DTpellet} = 1 \mathrm{x} 10^{19}$ m^{-3}/s, and the pellet injection period is 1 s (f_{pel}= 1 Hz). The summed value of continuous gas puffing and pellet fueling rate is shown in Figure 13 (e). The slowly time varying gas puffing can suppress the fusion power and density oscillations as demonstrated in the helical reactor design [28].

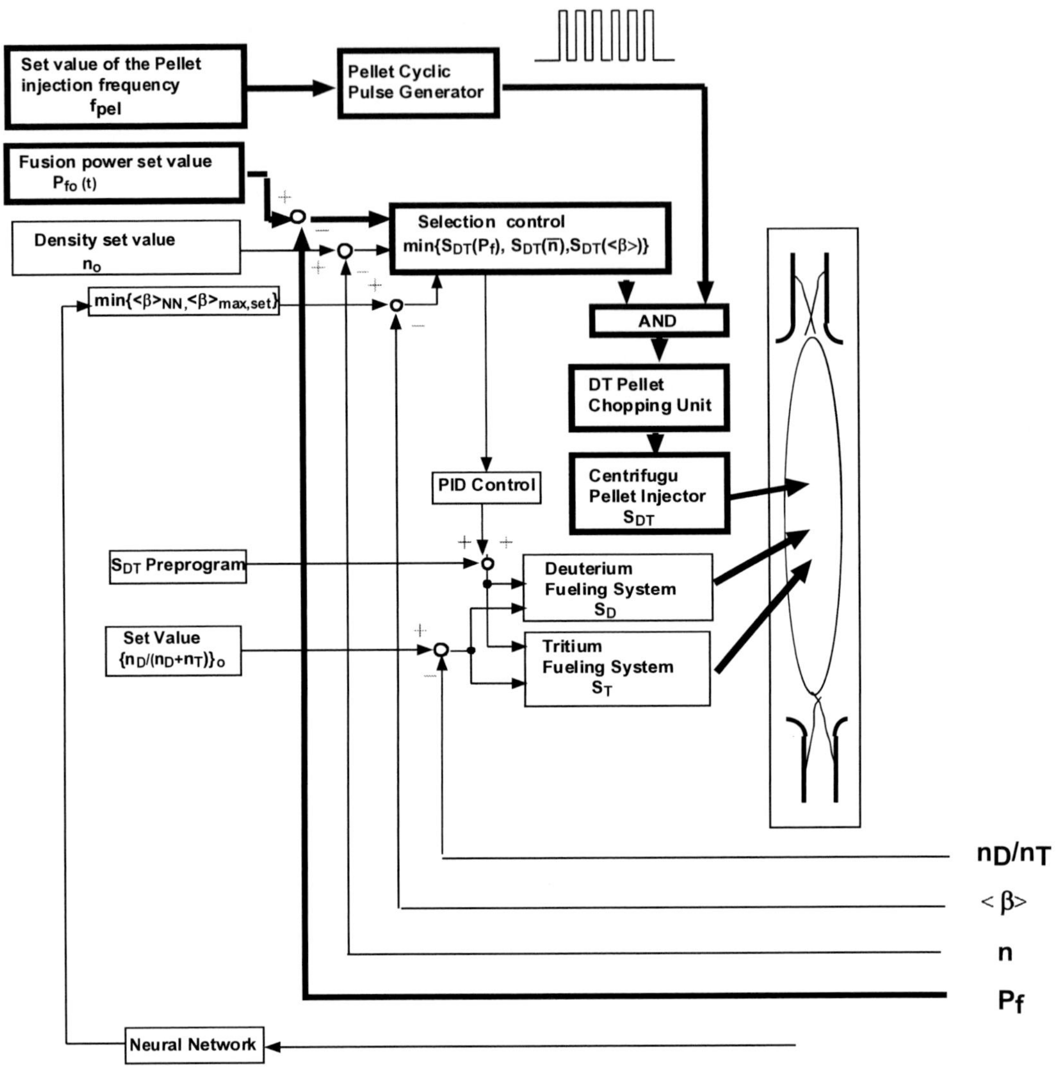

Figure 12. Pellet injection control diagram in addition to gas fueling using the centrifugal pellet injector.

5. Diagnostic Network System for Backing up Output Signals

Ignited operation is much safer against the diagnostic failure by employing the fail-safe control algorithm based on the fusion power, the density, and the beta measurements as described in section 3. To improve this situation further, various diagnostic systems are connected in a network as shown in Figure 14. The malfunction of some diagnostics can be properly compensated by the other working diagnostics. Detection method of the diagnostic malfunction should be incorporated in a system using such as a neural network.

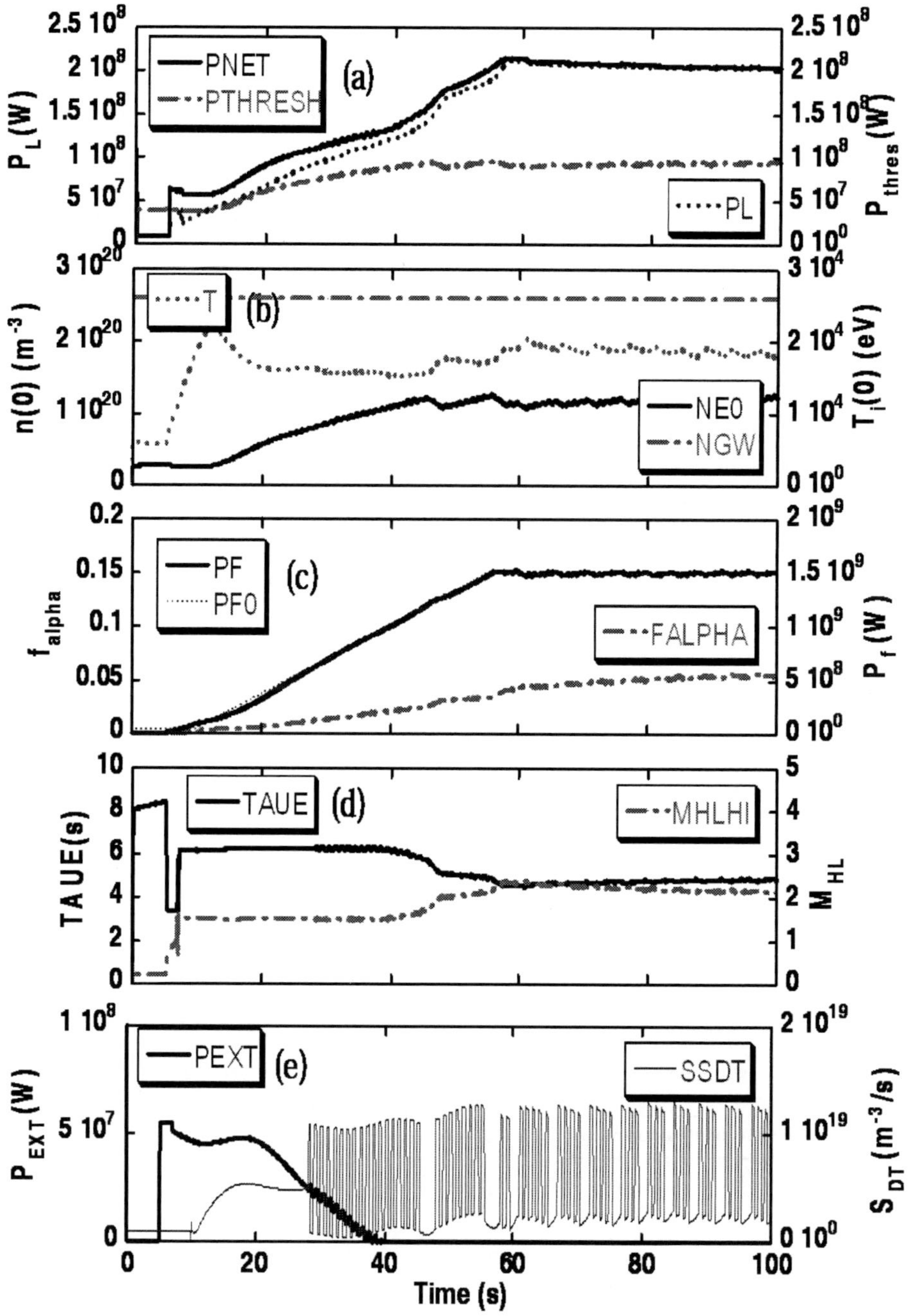

Figure 13. Ignited operation by pellet injection with continuous gas puffing for n_{GW}=2.1x10^{20} m^{-3} and $<\beta>_{max}$ = 8.0 %. Pellets were injected when the density reaches 8x10^{19} m^{-3}. The summed fueling rate from pellet and gas puffing is shown in (e). The confinement time scaling is IPB98(y,2) scaling with γ_{HH}=1.0 and the H-mode power threshold given by Righi et al [22].

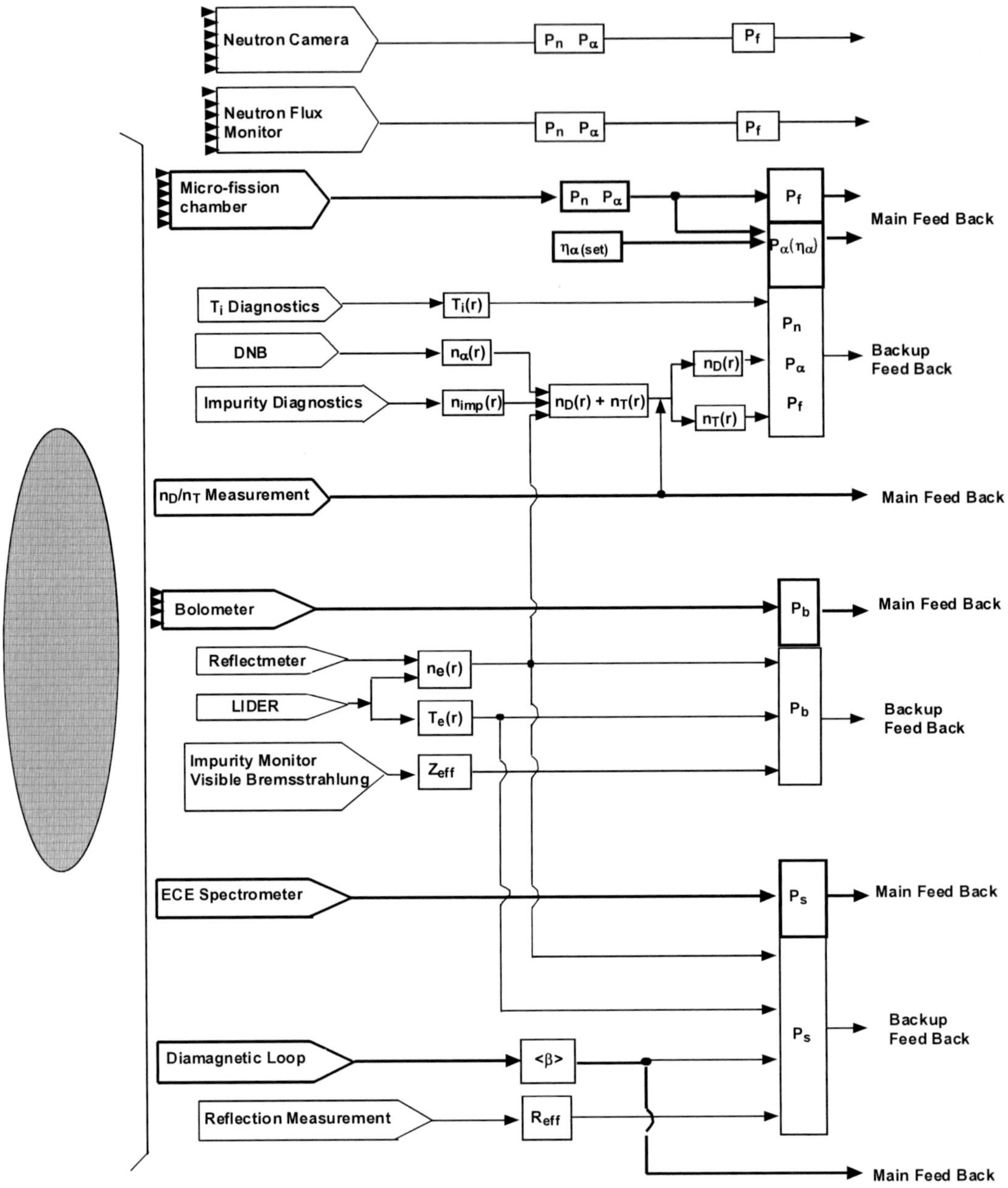

Figure 14. Possible diagnostic network system for backing up diagnostic system for ignition feedback control.

6. Further Issues on Control

We believe that the control diagram in Figure 10 shows the essential part of the core ignition burn control, although more detailed algorithm should be developed. Various interlock systems to protect machine such as disruption avoidance techniques etc should be added to this control algorithm for actual design of the control algorithm. The connection between the heating equipment and the calculated heating power should be disconnected in abnormal events and then plasma current should be shut down by the killer pellet injection [28]. Divertor control using the impurity injection should be also implemented as seen in Figure 1 in the reference [18], which needs more study.

Although the current profile control is effective to reduce the sawtooth crashes even when the inductive operation, the plasma current profile and confinement were assumed to be decoupled in the ignition burn in the inductive operation. Current profile control is most effective for an advanced tokamak (AT) operation with the reversed shear current profile to have a large confinement factor and large bootstrap current fraction. The current profile may determine the density profile and temperature profiles through control of the thermal conductivity profile, and the pressure profile also determines the current profile through the bootstrap current. Therefore, the self-consistent treatment is important for AT operation control. The basic ignition control algorithm described in this paper should be somehow incorporated into the control algorithm for AT operation. Control algorithm of the external heating power would be replaced to have the AT operation instead of the H-mode operation. (We note that actually in a helical reactor the density limit depending on the heating power has been used for the external heating power control [28].) As an AT operation is essentially a sub-ignited operation by the non-inductive current drive power, the operating point is far from the ignition boundary and may not reach the thermally unstable ignition boundary. Thus we notice that an AT operation is inherently in the thermally stable regime different from the usual ignited operation in an inductive operation.

Electron cyclotron current drive (ECCD) power would be applied to suppress the neo-classical tearing mode (NTM) which is excited when the safety factor q profile evolves at the low order rational values. This excess current drive power would be flown into the divertor regime, or the other heating power needs to be reduced to keep the total power. As the resistive wall mode (RWM) may limit the maximum beta value, feedback control system using the non-axisymmetric coils to alter the plasma rotation would be incorporated to provide the set value of the maximum beta in the basic ignition control algorithm.

Although some theoretical work on the plasma current ramp-up by the lower hybrid current drive, ICRH and NBI with the current profile control has been done [31] in the condition of no central solenoid, the poloidal coil effect on the plasma current ramp-up has not been taken into account. Recent JT-60U experiments on the plasma current ramp-up demonstrate its importance of the inductive loop voltage on the current ramp-up from the vertical field coil [32]. This means that the traditional calculation on the current ramp-up by the non-inductively driven current alone is not enough for control of the current profile. In a similar work on the spherical tokamak reactor without central solenoid, the vertical field created by the outer vertical field coils changes the plasma current when the plasma energy is changed by disturbance [33]. Therefore, the disturbance during discharge should be reduced as small as possible in AT operation when a central solenoid is not activated.

In this review article, developments of the global ignition burn control algorithm for an inductive operation in a tokamak reactor has been presented. Although the AT operation is sub-ignition, the burn control presented here would work in the sub-ignited operation with the non-inductive current drive [34] and then be combined with the profile control for AT operation.

APPENDIX

We present the 0-dimenstional particle and power balance equations here. We assume the following density profiles: $n_e(x)/n_e(0) = n_D(x)/n_D(0) = n_T(x)/n_T(0) = n_\alpha(x)/n_\alpha(0) = (1 - x^2)^{\alpha_n}$. The deuterium and tritium particle balance equations are

$$\frac{dn_D(0)}{dt} = (1+\alpha_n) S_D(t) - (1+\alpha_n) n_D(0) n_T(0) \overline{<\sigma v>_{DT}(x)} - \frac{n_D(0)}{\tau_D^*} \quad \text{(A-1)}$$

$$\frac{dn_T(0)}{dt} = (1+\alpha_n) S_T(t) - (1+\alpha_n) n_D(0) n_T(0) \overline{<\sigma v>_{DT}(x)} - \frac{n_T(0)}{\tau_T^*} \quad \text{(A-2)}$$

where $\overline{<\sigma v>_{DT}(x)}$ is the D-T fusion rate, $S_D(t)$ and $S_T(t)$ are the deuterium and tritium external fueling rates, respectively. $\tau_D^* = \tau_D/(1 - R_D)$ and $\tau_T^* = \tau_T/(1 - R_T)$ are the effective particle confinement time for deuterium and tritium with R_D and R_T being the recycling coefficients of the deuterium and tritium, respectively, and the bar denotes the volume averaged terms.

The peak alpha ash density $n_\alpha(0)$ is described using $\bar{n}_\alpha(x) = n_\alpha(0)/(1 + \alpha_n)$ as

$$\frac{dn_\alpha(0)}{dt} = (1+\alpha_n) n_D(0)\, n_T(0) \overline{<\sigma v>_{DT}(x)} - \frac{n_\alpha(0)}{\tau_\alpha^*} \quad \text{(A-3)}$$

where τ_α^* is the effective alpha ash confinement time which is assumed to be constant radially, and is given by $\tau_\alpha^* = \tau_\alpha/(1 - R_\alpha)$ with τ_α the alpha ash confinement time and R_α the recycling coefficient of the alpha particle on the first wall. The electron density is calculated by the charge neutrality condition:

$$n_e(0) = n_D(0) + n_T(0) + 2n_\alpha(0) + Zn_{imp} \quad \text{(A-4)}$$

where $\bar{n}_e(x) = n_e(0)/(1 + \alpha_n)$ and $Zn_{imp} = Z_{Ar}n_{Ar} + Z_{Be}n_{Be}$ with Z_{Ar} being the ion charge of Argon, Z_{Be} the ion charge of Beryllium, and n_{Ar} the argon density, and n_{Be} the Beryllium density. With the flat impurity profile $n_{imp} = f_{imp}n_e(0)$ and $n_{Be} = f_{Be}n_e(0)$, where the impurity fractions f_{imp} and f_{Be} are determined by the central electron density, the electron density is rewritten by

$$n_e(0) = \frac{1}{\{1 - (1+\alpha_n)Zf_{imp}\}} \{n_D(0) + n_T(0) + 2n_\alpha(0)\} \quad \text{(A-5)}$$

where $Zf_{imp} = Z_{Ar}f_{Ar} + Z_{Be}f_{Be}$. Here, the effective ion charge is $Z_{eff} = f_D + f_T + 4 f_\alpha + Z_{Be}^2 f_{Be} + Z_{Ar}^2 f_{Ar}$.

The time dependent power balance equation for the peaked value is

$$\frac{dT_i(0)}{dt} = \frac{1+\alpha_n+\alpha_T}{1.5\,e\left(f_D+f_T+1/\gamma_i+f_\alpha\right)n_e(0)}\left[P_{EXT}/V_o+\bar{P}_\alpha+\bar{P}_{oh}-\left\{\bar{P}_L+\bar{P}_b+\bar{P}_s\right\}\right]$$

$$-\frac{T_i(0)}{\left(f_D+f_T+1/\gamma_i+f_\alpha\right)n_e(0)}\left[\left\{1+\frac{1}{\gamma_i\left(1-(1+\alpha_n)Z\,f_{imp}\right)}\right\}\frac{dn_D(0)}{dt}\right.$$

$$\left.+\left\{1+\frac{1}{\gamma_i\left(1-(1+\alpha_n)Z\,f_{imp}\right)}\right\}\frac{dn_T(0)}{dt}+\left\{1+\frac{2}{\gamma_i\left(1-(1+\alpha_n)Z\,f_{imp}\right)}\right\}\frac{dn_\alpha(0)}{dt}\right] \quad \text{(A-6)}$$

where the temperature ratio is $\gamma_i = T_i(0)/T_e(0)$, the deuterium density fraction is $f_D = n_D(0)/n_e(0)$, the tritium density fraction is $f_T = n_T(0)/n_e(0)$, the alpha ash density fraction is $f_\alpha = n_\alpha(0)/n_e(0)$ and $e = 1.6\text{x}10^{-19}$. Thus, Eqs (A-1), (A-2), (A-3) and (A-6) are solved simultaneously, and the electron density is calculated by Eq. (A-5). Here V_o is the plasma volume, P_{EXT} is the external heating power, $\bar{P}_{oh}$ is the Ohmic heating power density, $\bar{P}_L$ is the plasma conduction loss per unit volume, $\bar{P}_\alpha$ is the alpha heating power density, $\bar{P}_b$ is the bremsstrahlung loss per unit volume, and $\bar{P}_s$ is the synchrotron radiation loss per unit volume.

Each term in the power balance equation in Eq. (A-6) is described as follows:

[1] The plasma conduction loss is $P_L = 1.5\ (f_D+f_T+1/\gamma_i+f_\alpha)T_i(0)\ 1.6\ \text{x}10^{-19}/(1+\alpha_n+\alpha_T)n_e(0)/\tau_E$

[2] The alpha heating power is $P_\alpha = f_D\, f_T\, (3.52\,\eta_\alpha)\overline{<\sigma v>_{DT}(x)}\,1.6\text{x}10^{-13}\, n_e(0)^2$, η_α being the alpha heating efficiency.

[3] The ohmic heating power is P_{oh} including the neoclassical resistivity.

[4] The bremsstrahlung loss is $P_b = A_b n_e(0)^2$ which includes the ion-electron and electron-electron scattering with the relativistic effects. The coefficient A_b is given by

$$A_b=1.5\times10^{-38}Z_{eff}\left[\frac{1}{1+2\alpha_n+0.5\alpha_T}+\frac{1}{1+2\alpha_n+1.5\alpha_T}\cdot\frac{2T_e(0)}{mc^2}\right]\sqrt{T_e(0)}$$
$$+3\times10^{-38}\left[\frac{1}{1+2\alpha_n+1.5\alpha_T}+\frac{0.5}{1+2\alpha_n+2.5\alpha_T}\cdot\left(\frac{T_e(0)}{mc^2}\right)\right.$$
$$\left.-\frac{3}{1+2\alpha_n+3.5\alpha_T}\cdot\left(\frac{T_e(0)}{mc^2}\right)^2\right]\frac{2T_e(0)^{1.5}}{mc^2} \quad \text{(A-7)}$$

where mc^2 is the electron rest energy.

[5] The synchrotron radiation loss is $P_s = A_s n_e(0)^2$ which includes the relativistic effects, the torus effects, and the non-reflecting surfaces such as holes. The coefficient A_s is given by

$$A_s = 2.5\times10^{-56} T_e(0)^4 \gamma_i^{1.5} \frac{\{f_D + f_T + 1/\gamma_i + f_\alpha\}^{1.5}}{\beta(0)^{1.5}\sqrt{aB_t}} \left\{ f_H + (1-f_H)\sqrt{1-R_{eff}} \right\}$$
$$\times \int_0^1 \left[(1-x^2)^{0.5\alpha_n + 2.5\alpha_T} \left\{ 1 + T_e(0)(1-x^2)^{\alpha_T}/204000 \right\} \right.$$
$$\left. \times \left\{ 1 - \beta(0)(1-x^2)^{\alpha_n+\alpha_T} \right\}^{5/4} \sqrt{\sqrt{T_e(0)}(1-x^2)^{0.5\alpha_T} + \frac{2a}{R}\sqrt{\frac{mc^2}{2\pi}}} \right] 2x dx \qquad \text{(A-8)}$$

where R_{eff} is the wall reflectivity, f_H is the hole fraction, the central toroidal beta is given by $\beta(0) = <\beta>(1 + \alpha_n + \alpha_T)$ with the average toroidal beta $<\beta> = (f_D + f_T + 1/\gamma_i + f_\alpha) n_e(0)T_i(0)/\{(1 + \alpha_n + \alpha_T)(B_o^2/2\mu_o)\}$.

The ITER93HP scaling [19] is

$$\begin{cases} \tau_{ITER93HP}[s] = 0.0532 A_i^{0.41} I_p^{1.06}[MA]\, \bar{n}_{20}^{0.17}[\times10^{20}m^{-3}]\, R^{1.9}[m]\, a^{-0.11}[m] \\ \qquad \times \kappa^{0.66} B_t^{0.32}[T]/P_{HT}^{0.67}[MW] \\ \tau_{AUX} = \gamma_{HH} 0.85 \times \tau_{ITER93HP} \end{cases} \qquad \text{(A-9)}$$

and the IPB98 (y,2) confinement time scaling [20] is

$$\begin{cases} \tau_{IPB98(y,2)} = 0.0562 A_i^{0.19} I_p^{0.93}[MA]\, \bar{n}_{19}^{0.41}[\times10^{19}m^{-3}]\, R_o^{1.97}[m]\, \varepsilon \\ \qquad \times \kappa^{0.78} B_{to}^{0.15}[T]/P_{HT}^{0.69}[MW] \\ \tau_{AUX} = \gamma_{HH}\tau_{IPB98(y,2)} \end{cases} \qquad \text{(A-10)}$$

The final confinement time τ_E is given by

$$\tau_E = \min\{\tau_{NA}, \tau_{\square\square\square}) \qquad \text{(A-11)}$$

with the Neo-Alcator scaling :

$$\tau_{NA}(s) = 7.1\times10^{-22}\, \bar{n}[cm^{-3}]\, R^{2.04}[cm]\, a^{1.04}[cm]\sqrt{q_a} \qquad \text{(A-12)}$$

The second H-mode threshold power is given by [22]

$$\bar{P}_{thresh}[MW] = 2.84\bar{n}^{0.58}[10^{20}m^{-3}]B_{to}^{0.82}[T]R_o^{1.0}[m]a^{0.81}[m]/A_i \qquad \text{(A-13)}$$

REFERENCES

[1] Rebut, P-H et al., in *Plasma Physics and Controlled Nuclear Fusion Research*, (15h International Conference, Seville, Spain, 1994), Vol. 2, IAEA, Vienna, (1995) 451.
[2] Shimomura, Y., *Phys. Plasmas*, 1 (5) (1994) 1612.
[3] Aymar, R., Chuyanov, V., Huguet, M., et al., in 16th IAEA *Fusion Energy Conference*, (Montréal, Canada, 7-11 October, 1996) Vol .1, IAEA, Vienna (1997) 3.
[4] Budny, R.V., McCune, D.C., Redi, M.H., et al., *Phys. Plasmas*, 3 (12) (1996) 4583.
[5] ITER-JCT Team and Home Team (presented by Janeshitz G), *Plasma Physics and Controlled Fusion*, 37 (1995) A19.
[6] Boucher, D. and Makhovatov. V., *Plasma Physics & Controlled Fusion*, 38 (1996) 1225.
[7] Wesley. J. C., and Iter Joint Central Team, *Phys. Plasmas* 4 (7) (1997) 2642.
[8] Costley, A. E., et al., "Requirements for ITER Diagnostics", p23, in "Diagnostics for Experimental Thermonuclear Fusion Reactors", *Proceedings of the International Workshop on diagnostics for ITER*, (Aug. 28-Sep. 1, 1995, Varenna, Italy), Ed. by P. E. Stott, G. Gorini and E. Sindoni, Plenum Press, New York, 1996.
[9] Mukhovatov. V. S., et al., "Role and Requirements for Plasma Measurements on ITER", p25, in "Diagnostics for Experimental Thermonuclear Fusion Reactors 2", *Proceedings of the International Workshop on diagnostics for ITER*, (Sep. 4-12, 1997, Varenna, Italy), Ed. by P. E. Stott, G. Gorini, P. Prandoni, and E. Sibdoni, Plenum Press, New York and London, 1998.
[10] Costley, A., et al., "Overview of the ITER Diagnostic System", p41, in "Diagnostics for Experimental Thermonuclear Fusion Reactors 2", *Proceedings of the International Workshop on diagnostics for ITER*, (Sep. 4-12, 1997, Varenna, Italy), Ed. by P. E. Stott, G. Gorini, P. Prandoni, and E. Sibdoni, Plenum Press, New York and London, 1998.
[11] Mitarai, O. and Muraoka, K., *Plasma Physics & Controlled Fusion*, 36 (1996) 551.
[12] Mitarai, O. and Muraoka, K., *Nucl. Fusion*, 37, (1997) 1523.
[13] Mitarai, O. and Muraoka, K., *Plasma Physics & Controlled Fusion*, 40 (1998) 1349.
[14] Schuster, E., Slavkrstic´, M., and Tunan, G., *Fusion Science and Technology*, 43 (2003) 18.
[15] Mitarai, O. and Muraoka, K., "Ignition Burn Control and Diagnostics in ITER", p93, in "Diagnostics for Experimental Thermonuclear Fusion Reactors 2", *Proceedings of the International Workshop on diagnostics for ITER*, (Sep. 4-12, 1997, Varenna, Italy), Ed. by P. E. Stott, G. Gorini, P. Prandoni, and E. Sibdoni, Plenum Press, New York and London, 1998.
[16] Mitarai, O. and Muraoka, K., *Nucl. Fusion*, 39,(1999) 725
[17] Mitarai, O. and Muraoka, K., , *Fusion Technology*, 36 (1999) 194.
[18] Mitarai, O. " Fuel Ratio and Fueling Control for Safe Ignited Operation in ITER class Tokamak Reactors", *Advances in Plasma Physics Research*, Vol 12,(2002, 2) p37-74, Ed by G Francois, Nova Science Pub
[19] Iter H-mode Database Working Group, *Nucl. Fusion*, 34, (1994) 131.
[20] "Iter Physics Basis", *Nucl. Fusion*, 39, (1999) 2137.
[21] Takizuka, T., Boucher,D. , et al., "Threshold Power and Energy Confinement for ITER", in 16th IAEA Fusion Energy Conference, (Montreal, Canada, 7-11 October, 1996), IAEA-CN-64/F-5, (International Atomic Energy Agency, Vienna), Vol.2, p795.

[22] Righi, E., Bartlett, D. V., Christensen, J. P., Conway, G. D., et al., *Nucl. Fusion*, 39, (1999) 309.
[23] Johnson. L. C., et al.,"Overview of Proton Product Diagnostics for ITER", p409, in "Diagnostics for Experimental Thermonuclear Fusion Reactors 2", *Proceedings of the International Workshop on diagnostics for ITER*, (Sep. 4-12, 1997, Varenna, Italy), Ed. by P. E. Stott, G. Gorini, P. Prandoni, and E. Sibdoni, Plenum Press, New York and London, 1998.
[24] Marcus. F.B., et. al., "A Radial Neutron Camera Design for ITER with Integral Shielding", p419, in "Diagnostics for Experimental Thermonuclear Fusion Reactors 2", *Proceedings of the International Workshop on diagnostics for ITER*, (Sep. 4-12, 1997, Varenna, Italy), Ed. by P. E. Stott, G. Gorini, P. Prandoni, and E. Sibdoni, Plenum Press, New York and London, 1998.
[25] Wurden, G. A., "Imaging Bolometry Development for Large Fusion Devices", p399, in "Diagnostics for Experimental Thermonuclear Fusion Reactors 2", *Proceedings of the International Workshop on diagnostics for ITER*, (Sep. 4-12, 1997, Varenna, Italy), Ed. by P. E. Stott, G. Gorini, P. Prandoni, and E. Sibdoni, Plenum Press, New York and London, 1998.
[26] Mitarai, O., Honda, T. , Nishitani, T., and Muraoka, K., *Fusion Engineering and Design*, 55, (2001) 477-500.
[27] Greenwald, M., et al., *Nucl. Fusion*, 28 (1988) 2199
[28] Mitarai, O., Oda, A., Sagara, A., Yamazaki, K., and Motojima, O., *Fusion Engineering and Design*, 70, (2004) 247-267
[29] Viniar, I., Sudo, S., Geraud, A., et al., *Fusion Engineering and Design*, 56-59 (2001) 295.
[30] Yoshino, R., *Plasma Physics and Controlled Fusion*, 4 (1997) 313
[31] Moreau, D., and Voitsenkhovich, I., *Nucl. Fusion*, 39 (1999) 685
[32] Shiraiwa, S., Takase, Y., Fukuda, T., Gao, X., Gryaznevich, M., Ide, S., Itoh, S., Kamada, Y., Maekawa, T., Mitarai, O., Miura, Y., ,et al. *Physical Rev. Lett.*, Vol.92 Number 3, (2003) p035001-1
[33] Mitarai, O., and Takase, Y., *Fusion Science and Technology*, 43, No. 1 (2003) 67.
[34] Mitarai, O., Yoshino, R., and Ushigusa, K., *Nucl. Fusion* 10 (2002) 1257

In: Perspectives in Mathematical Physics
Editor: Gregory D. Monteggo
ISBN: 978-1-61122-984-4

ESTAFETTE OF DRIFT RESONANCES, STOCHASTICITY AND PARTICLE TRANSPORT

Alexander A. Shishkin
Institute of Plasma Physics, National Science Center "Kharkov Institute of Physics and Technology", Kharkov- 108, Ukraine,
and
Department of Physics and Technology, Kharkov "V.N. Karazin" National University, Kharkov-77, Ukraine

ABSTRACT

1. What is the Estafette of Resonances? The question can arise: what is the "estafette", what does it mean. This word had come from French to Russian and it is known from sport. If one runner with the stick in his hand starts to run and then some time later when one part of the path is already passed brings this stick to another runner, and this runner in his turn will bring the same stick to the following runner we say that this is ***estafette***. Such picture can help us to explain what is the estafette in physics, particularly in the physics of a particle transport. In this case the test particle is the "stick" and the adjacent drift resonances are the "runners". If such adjacent resonances are spread in the space, for example, along the radial direction of plasma cross-section, then the particle can be brought across the plasma - from the center to the periphery or vise versa i.e. from the periphery to center. How can we arrange and control these resonances. At first this terminology and this approach was proposed in paper [1], where the removal of the cold alpha-particles in fusion reactor was considered.

2. What is Drift Resonance? Drift resonance is that situation when there is the drift surface with the rational value of the drift rotational transform iota* = m/n, where "m" and "n" are the numbers, and there is the perturbation of the magnetic field with the "wave numbers" "m" and "n". In that case on the place of the rational surface the drift island occurs [2]. Drift islands can take place on the adjacent drift surfaces if they have the rational values of drift rotational transform iota*=m'/n', iota *=m''/n'' and so on and at the same moment there are the magnetic field perturbations with the corresponding sets of wave numbers (m' and n', m'' and n'' and so on).

3. The overlap of adjacent drift resonances can lead to stochasticity if the amplitudes of the perturbation filed are rather large and the distance between the adjacent resonances is rather small [1].

4. The principal NEW thing here is that perturbations of the magnetic field are being "switched on" consequently one after another. Namely this thing can lead to estafette of resonances and particle transport across the magnetic surface.

5. Now we know that the estafette of drift resonances can take place in the plasma with the ambipolar electric field taken into account. It means that more wide physics

phenomenon, particularly Alfven type perturbation and particle interaction in plasma can be understood and described with the use of estafette of resonance.

6. Practical usefulness of the estafette of drift resonances is the ***removal of cold alpha particles and impurity ions*** from the plasma core. Applicability can be realized with the use of cooper coils on the as there are on Large Helical Device [3,4]. Another very important applicability is that estafette of resonances can be the physics mechanism for the ***interaction of MHD magnetic field perturbations & particles*** which can lead to the losses of the energetic particles from plasma.

1. Motivation of Study

A new method of particle motion control in toroidal magnetic traps with rotational transform using the *estafette* of drift resonances and stochasticity of particle trajectories is proposed. The use of the word "estafette" here means that the particle passes through a set of resonances in consecutive order from one to another during its motion. The overlapping of adjacent resonances can be moved radially from the center to the edge of the plasma by switching on the corresponding perturbations in accordance with a particular rule in time. In this way particles (e.g. cold alpha-particle) can be removed from the center of the confinement volume to the plasma periphery. For the analytical treatment of the stochastic behaviour of particle motion the stochastic diffusion coefficients $D_{r,r}, D_{r,\vartheta}, D_{\vartheta,\vartheta}$ are introduced. The new approach is demonstrated by numerical computations of the test helium particle trajectories in the toroidal trap Large Helical Device.

Different tasks in fusion need the control of particle motion. Among them are ***the removal of the "cold" alpha particles (helium ash) from a fusion reactor*** and in simulation experiment in modern devices, the conversion of trapped particles into passing ones to reduce the neoclassical transport , the screening of the bulk plasma from impurity ions and others. There exist different approaches to control the particle motion and one of them is to apply magnetic field perturbations to produce resonance trajectories - drift islands, overlapping of the islands and stochastic trajectories that lead, for example, to the removal of test particles from the center of the magnetic configuration to the periphery [4,5]. The drift island motion – without an overlapping of resonances is possible [2,3,6].

In this paper a new approach is considered, namely the *estafette* of resonances for the particle. Here the word *estafette* (relay race) is used to describe that the particle passes through a set of resonances consistently from one to another during its motion. A new method of particle motion control in toroidal magnetic traps with rotational transform using the *estafette* of drift resonances and stochasticity of particle trajectories can be described in the following. If there are some adjacent rational drift surfaces with the drift rotational transform $\iota^* = n/m, \iota^* = n'/m'$, $\iota^* = n''/m''$, then the magnetic perturbations with the wave numbers (m,n), (m',n'), (m'',n'') can lead to some families of drift islands. Overlapping of the adjacent resonance structure is the reason for the stochasticity [7] of the particle trajectories. If a particle trajectory passes through the set of perturbations this test particle can escape from the center of the confinement volume to the periphery.

Overlapping of the adjacent resonances in the local space within the plasma can be transferred outward in the process of particle motion by switching on the corresponding perturbations in accordance with a special rule in time. In such a way the particle (cold alpha-particle, for example) can be removed from the center of the confinement volume into the periphery of the plasma and the "reverse" task - the transport of the particles injected from outside the plasma into the center of the confinement volume – can be solved.

The mathe necessary for the deviation of the particle trajectory from the initial surface in a certain angular space under the given deviation in the radial direction. This approach can be used in a divertor flux description and for the optimization of the divertor configuration. Stochastic diffusion coefficients for magnetic field lines were introduced in [10,11]. The concept of braided magnetic lines is very fruitful to describe the transport of plasma across destroyed magnetic surfaces [12-15].

In this paper the new physics effect – *estafette* of drift resonances of the test particle (for instance "cold" alpha-particles) in the toroidal trap with a rotational transform of the magnetic field - here the Large Helical Device is used as an example - and the removal of test particles from the center of the magnetic configuration to the periphery is shown by the numerical integration of guiding center equations. The numerical study is supported by the analytical development

The method of the *estafette* of resonances can be applied to other practical physics tasks and may be considered as a new approach in the transition to chaotic states [16].

The paper is organized as follows. The main equations are described in Section 2: the guiding center equations are in the subsection 2.1 and the main magnetic field – in sub section 2.2 and perturbations of the magnetic field in subsection 2.3. The analytical treatment is given in Section 3, where the diffusion type equation for the drift surface function is obtained (subsection 3.1) and stochastic diffusion coefficients are introduced (subsection 3.2). *Estafette* of resonances of the test particle in a real magnetic configuration with the numerical integration of particle guiding center equations is analyzed in Section 4. Coil system for the producing the perturbations, the physics mechanisms which can effect the resonance conditions are discussed in Section 5. Main conclusions are given in Section 6.

2. Main Equations

2.1. Guiding Center Equations

For our consideration guiding center equations are used [17]

$$\frac{d\mathbf{r}}{dt} = V_{II}\frac{\mathbf{B}}{B} + \frac{c}{B^2}[\mathbf{E}\times\mathbf{B}] + \frac{M_j c(2V_{II}^2 + V_\perp^2)}{2eB^3}[\mathbf{B}\times\nabla B],$$

$$\frac{dW}{dt} = e\mathbf{E}\frac{d\mathbf{r}}{dt} + \frac{M_j V_\perp^2}{2B}\frac{\partial B}{\partial t}, \qquad (2.1)$$

$$\frac{d\mu}{dt} = 0.$$

Here **r** is the radius-vector of the particle, **B** is the magnetic field, **E** is the electric field, $V_{\parallel}$ and $V_{\perp}$ are the parallel and perpendicular velocities, M_j and e are mass and charge of the particle, W is the kinetic energy and μ is the magnetic moment of the particle ($\mu = \dfrac{M_j V_{\perp}^2}{2B}$).

2.2. Main Magnetic Field

The main magnetic field ($\mathbf{B} = \nabla\Phi$) is modeled with the use of the magnetic field potential

$$\Phi = B_0 \left[R\varphi - \frac{R}{m} \sum_n \varepsilon_{n,m} (r/a_h)^n \sin(n\vartheta - m\varphi) + \varepsilon_{1,0} r \sin\vartheta \right], \qquad (2.2)$$

where B_0 is the magnetic field at the circular axis, R and a_h are the major and minor radii of the helical winding; r, ϑ, φ are the coordinates connected with the circular axis of the torus, r is the radial variable, ϑ and φ are the angular variables along the minor and major circumference of the torus, ϑ increases in the direction opposite to the main normal of the circular axis of the torus; metric coefficients are the following: $h_r = 1$, $h_\vartheta = r$, $h_\varphi = R + r\cos\vartheta$; m is the number of the magnetic field periods along the torus, l is the helical winding pole number. The index n assumes the values $n = l$, $l-1$, $l+1$; $\varepsilon_{n,m}$ are the coefficients of the harmonics of the magnetic field. For the results presented here the parameters of the Large Helical Device [18] are taken: $l = 2$, $m = 10$, $B_0 = 3$ T, $R = 390$ cm, $a_h = 97.5$ cm; the values of $\varepsilon_{n,m}$ are taken in such a way that the magnetic surfaces, the magnetic field modulation along the force line and other properties coincide with the results of Ref. [19]. For the configuration with the inward shift of the magnetic axis the parameters are $\varepsilon_{2,10} = 0.79$, $\varepsilon_{3,10} = -0.032$, $\varepsilon_{1,10} = -0.056$, $\varepsilon_{1,0} = 0.045$.

2.3. The Perturbation Magnetic Field

The perturbation magnetic field considered here is given with the scalar potential

$$\Phi_p = B_0 a_h \frac{\varepsilon_{m,n,p}}{m_p} (r/a_h)^{m_p} \sin(m_p\vartheta - n_p\varphi + \delta_{m,n,p}) \qquad (2.3)$$

The “wave numbers”, amplitude and phase values of the perturbing magnetic field should be chosen in accordance with the rational values of the drift rotational transform of the resonant particle. In the case considered below these parameters are the following: m_p=4, n_p=3; $\varepsilon_{4,3,p}$= 0.003, $\delta_{4,3,p} = \pi/2$; m_p= 5, n_p=4 $\varepsilon_{5,4,p}$=0.007, $\delta_{5,4,p} = 0$; m_p= 10, n_p=9 $\varepsilon_{9,10,p}$=0.03, $\delta_{9,10,p} = 0$; all explanations can be found in Section 4.2.

3. Analytical Treatment

3.1. Equation for the Drift Flux Surface Function

As is known [20, 21], particle motion can be described with the integrals of the equation system (2.1) and one of them is the function of the drift surface $\Psi^*(r,\vartheta,\varphi,V_\parallel,V_\perp,t)$. We assume that the flux of particle guiding centers is conserved during the particle motion in analogy with the magnetic flux. This leads to

$$\Psi^*(r,\vartheta,\varphi,V_\parallel,V_\perp,t) = const \tag{3.1}$$

and the total derivative of the drift surface function is equal to zero

$$\frac{d\Psi^*}{dt} = 0 \tag{3.2}$$

The equation (3.2) in variables r,ϑ,φ and $V_\parallel$, $V_\perp$ takes the form

$$\frac{\partial\Psi^*}{\partial t} + \frac{dr}{dt}\frac{\partial\Psi^*}{\partial r} + \frac{rd\vartheta}{dt}\frac{\partial\Psi^*}{r\partial\vartheta} + \frac{Rd\varphi}{dt}\frac{\partial\Psi^*}{R\partial\varphi} + \frac{dV_\parallel}{dt}\frac{\partial\Psi^*}{\partial V_\parallel} + \frac{dV_\perp}{dt}\frac{\partial\Psi^*}{\partial V_\perp} = 0 \tag{3.3}$$

After substituting the equations for $\frac{dr}{dt}, \frac{rd\vartheta}{dt}, \frac{Rd\varphi}{dt}$ and $\frac{dV_\parallel}{dt}, \frac{dV_\perp}{dt}$ from equation system (2.1) into (3.3) we obtain

$$\frac{\partial\Psi^*}{\partial t} + \frac{V_\parallel}{B}(\mathbf{B}\nabla\Psi^*) + \frac{Mc}{2eB^3}\left(2V_\parallel^2 + V_\perp^2\right)([\mathbf{B}\times\nabla B]\nabla\Psi^*) - \frac{V_\perp^2}{B^2}(\mathbf{B}\nabla B)\frac{\partial\Psi^*}{\partial V_\parallel} + \frac{V_\parallel}{4B^2}(\mathbf{B}\nabla B)\frac{\partial\Psi^*}{\partial V_\perp^2} = 0 \tag{3.4}$$

In our following consideration we assume that $\frac{\partial\Psi^*}{\partial V_\parallel} = 0$ and $\frac{\partial\Psi^*}{\partial V_\perp^2} = 0$. These derivatives are very important when the transition of a particle from a trapped one into a passing one is studied. However they are not so important for the study of the resonant passing particle in the narrow range of $V_{\|}$ and $V_{\perp}$ especially, when the spreading of $\Psi^* = const$ contours due to magnetic field perturbations is considered.

We take here the simplified form of the magnetic field with the rotational transform $\iota(r^2)$

$$\mathbf{B}_0 = B_0\left\{0, \frac{r}{R}\iota(r^2), \frac{1}{1+(r/R)\cos\vartheta}\right\} \tag{3.5}$$

and for the perturbations

$$\mathbf{B}_{m,n} = B_0\{b_{m,n}(r/a)^{m-1}\sin(m\vartheta - n\varphi + \delta_{m,n}), b_{m,n}(r/a)^{m-1}\cos(m\vartheta - n\varphi + \delta_{m,n}), 0)\} \tag{3.6}$$

and

$$\widetilde{\mathbf{B}} = B_0\{b_{m',n'}(r/a)^{m'-1}\sin(m'\vartheta - n'\varphi + \delta_{m',n'}), b_{m',n'}(r/a)^{m'-1}\cos(m'\vartheta - n'\varphi + \delta_{m',n'}), 0)\} \tag{3.7}$$

Then the equation for $\Psi^*_{0,m,n}$ can be reduced to the following form

$$\frac{\partial \Psi^*_{0,m.n}}{\partial t} + \frac{V_{\|}}{B}\left(\mathbf{B}_{0,m,n} + \widetilde{\mathbf{B}}\right)\nabla\Psi^*_{0,m,n} + \frac{Mc}{2eB^3}\left(2V_{\|}^2 + V_{\perp}^2\right)\left[\left(\mathbf{B}_{0,m,n} + \widetilde{\mathbf{B}}\right)\times\nabla B\right]\nabla\Psi^*_{0,m,n} = 0 \tag{3.8}$$

where

$$\mathbf{B}_{0,m,n} = \mathbf{B}_0 + \mathbf{B}_{m,n} \tag{3.9}$$

We suppose that under the magnetic field $\mathbf{B}_{0.m.n}$ the function $\Psi^*_{0,m,n}$ is satisfied with the equation

$$\frac{V_{\|}}{B}\left(\mathbf{B}_{0,m,n}\nabla\Psi^*_{0,m,n}\right) + \frac{Mc}{eB^3}\left(2V_{\|}^2 + V_{\perp}^2\right)\left[\mathbf{B}_{0.m.n}\times\nabla B\right]\nabla\Psi^*_{0,m,n} = 0 \tag{3.10}$$

The function $\Psi^*_{0,m,n}$ is described with the expression [3]

$$\Psi^*_{0,m,n} = \int\left[m\iota(r/a)^2 - n\right]r\,dr + Rb_{m,n}\left(\frac{r}{a}\right)^{m-1} r\cos(m\vartheta - n\varphi + \delta_{m,n}) -$$

$$- \frac{McV_{\|}}{eB_0}\left(1 + \frac{V_{\perp}^2}{V_{\|}^2}\right)\frac{m\iota(r/a)^2 - n}{\iota(r/a)^2} r\cos\vartheta \tag{3.11}$$

Then under the effect of $\widetilde{\mathbf{B}}$ we find the change of $\Psi^*_{0,m,n}$ in the following way

$$\Psi^*_{0,m,n}\Big|_{\widetilde{\mathbf{B}}} = \Psi^*_{0,m,n}\Big|_{\widetilde{\mathbf{B}}=0} + \Delta r \frac{\partial \Psi^*_{0,m,n}}{\partial r} + r\Delta\vartheta \frac{\partial \Psi^*_{0,m,n}}{r\partial\vartheta} \tag{3.12}$$

Δr and $r\Delta\vartheta$ are due to $\widetilde{\mathbf{B}}$.

Then the equation (3.8) reduces to the following one

$$\frac{\partial \Psi^*_{0,m,n}}{\partial t} + \left(\widetilde{V}_r + \widetilde{V}_r \frac{\partial}{\partial r}\Delta r + \widetilde{V}_\vartheta \frac{\partial}{r\partial\vartheta}\Delta r\right)\frac{\partial \Psi^*_{0,m,n}}{\partial r} +$$

$$+\left(\widetilde{V}_\vartheta + \widetilde{V}_r \frac{\partial}{\partial r}(r\Delta\vartheta) + \widetilde{V}_\vartheta \frac{1}{r}\frac{\partial}{\partial\vartheta}(r\Delta\vartheta) - \widetilde{V}_r \frac{\Delta\vartheta}{r}\right)\frac{\partial \Psi^*_{0,m,n}}{r\partial\vartheta}$$

$$+ D_{r,r}\frac{\partial^2 \Psi^*_{0,m,n}}{\partial r^2} + D_{r,\vartheta}\frac{\partial^2 \Psi^*_{0,m,n}}{r\partial r\partial\vartheta} + D_{\vartheta,\vartheta}\frac{\partial^2 \Psi^*_{0,m,n}}{r^2\partial\vartheta^2} = 0 \tag{3.13}$$

Here such designations are introduced

$$\widetilde{\mathbf{V}} \equiv \frac{V_\parallel}{B}\widetilde{\mathbf{B}} + \frac{Mc}{2eB^3}\left(2V_\parallel^2 + V_\perp^2\right)\left[\widetilde{\mathbf{B}} \times \nabla B\right] \tag{3.14}$$

and

$$D_{r,r} \equiv \widetilde{V}_r \Delta r,\ D_{r,\vartheta} \equiv \widetilde{V}_r r\Delta\vartheta + \widetilde{V}_\vartheta \Delta r,\ D_{\vartheta,\vartheta} \equiv \widetilde{V}_\vartheta r\Delta\vartheta \tag{3.15}$$

3.2. Stochastic Diffusion Coefficients

Stochastic diffusion coefficients can be evaluated using

$$D_{r,r} = \widetilde{V}_r \Delta r \cong \left(\frac{V_\parallel}{B}\right)^2 (B_{m,n,r}\widetilde{B}_r + \widetilde{B}_r^2)\frac{2\pi R}{V},$$

$$D_{r,\vartheta} = \widetilde{V}_r r\Delta\vartheta + \widetilde{V}_\vartheta \Delta r \cong 2\left(\frac{V_\parallel}{B}\right)^2 (B_{m,n,r} + \widetilde{B}_r)(B_{m,n,\vartheta} + \widetilde{B}_\vartheta)\frac{2\pi R}{V},$$

$$D_{\vartheta,\vartheta} = \widetilde{V}_\vartheta r\Delta\vartheta \cong \left(\frac{V_\parallel}{B}\right)^2 (B_{m,n,\vartheta}\widetilde{B}_\vartheta + \widetilde{B}_\vartheta^2)\frac{2\pi R}{V}. \tag{3.16}$$

It should be noted that the sense of the diffusion coefficients is the following

$$D_{r,r} = \frac{(\delta r)^2}{\tau_{rr}}, \; D_{r,\vartheta} = \frac{r\delta r\delta\vartheta}{\tau_{r\vartheta}}, \; D_{\vartheta,\vartheta} = \frac{r^2(\delta\vartheta)^2}{\tau_{\vartheta\vartheta}} \tag{3.17}$$

From the expressions (3.17) (after substituting from (3.16)) we can evaluate, for example, the time τ_{rr} necessary for the deviation of particle trajectories from the initial surface in the radial direction, the time $\tau_{r\vartheta}$ necessary for the deviation of particle trajectories from the initial surface in a certain angular space under a given deviation in the radial direction and the time $\tau_{\vartheta\vartheta}$ for the deviation of the particle trajectories in the angular direction.

The stochastic diffusion coefficients obtained are the local characteristics and depend on the variables r, ϑ. We should remember that these diffusion coefficients describe the breaking of the separatrix that separates the m/n islands on the drift surface with the drift rotational transform $\iota^* = n/m$ from the adjacent resonance m'/n' islands on the drift surface with $\iota^* = n'/m'$ (Figure1). If we put the expressions for the magnetic field components (3.6) and (3.7) in (3.16) we can see that the stochastic diffusion coefficients depend on the variables r, ϑ.

One possible question can arise, namely, at which condition the drift islands can overlap and when the coefficients $D_{r,r}, D_{r,\vartheta}, D_{\vartheta,\vartheta}$ can be used.

As it is known the maximum transverse size of the drift island can be described by the following expression [3]

$$\Delta r_{n,m} = \left(\frac{2Rab_{n,m}(r_{n,m}/a)^m}{m(d\iota/dr^2)|}_{r=r_{n,m}} \right)^{1/2} \tag{3.18}$$

where $r_{n,m}$ is the radius of the magnetic surface with $\iota(r_{n,m}) = n/m$ and when the adjacent islands can overlap i.e. when [7]

$$K \equiv \frac{\Delta r_{n,m} + \Delta r_{n',m'}}{r_{n',m'} - r_{n,m}} \geq 1, (r_{n',m'} - r_{n,m} > 0) \tag{3.19}$$

then the stochasticity of the particle trajectory can occur.

Analytical expressions obtained in this paper can be used for the optimization of the magnetic field perturbations (the sets of "wave" numbers, amplitudes, phases) which can lead to stochastization of the test particle trajectories and the preservation of the drift surfaces of the bulk plasma particles. Obtaining such conditions is the subject of future work.

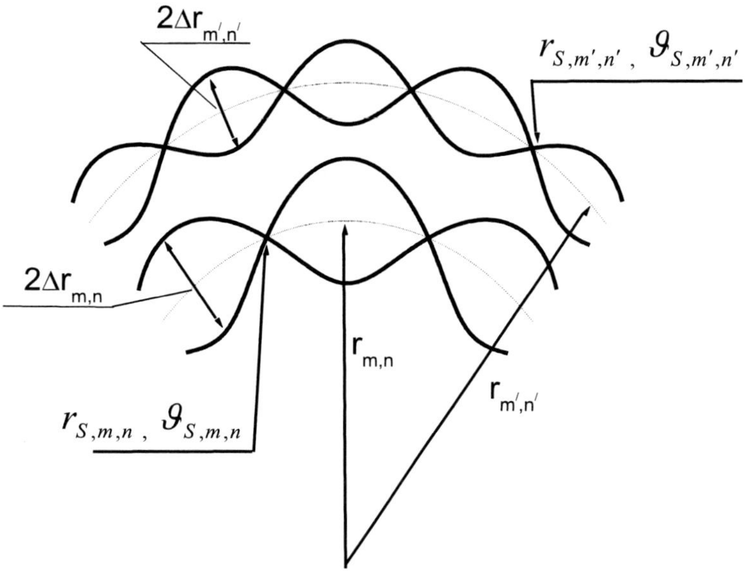

Figure 1. Part of two chains of islands: $r_{m,n}$ and $r_{m',n'}$ are the radii of the rational magnetic surfaces, $2\Delta r_{m,n}$ and $2\Delta r_{m',n'}$ are the widths of the islands, $r_{S,m,n}$, $\vartheta_{S,m,n}$ and $r_{S,m',n'}$, $\vartheta_{S,m',n'}$ are the coordinates of the separatrix ribs.

Here it is shown that under the perturbations with the "wave" numbers (m',n') the separartrix of the (m,n) resonance is broken and the particle trajectories become stochastic. Thus the particle can wander from one resonance (initial) to another resonance – adjacent to the initial one and then this phenomenon repeats, the particle passes step by step through the set of resonances and can escape from the center of the confinement volume to the periphery and vice versa from the periphery into the center of the confinement volume. In such a way the impurity ions (cold alpha-particles, for instance) can be removed from the magnetic trap. In the same way, the transport of high energy particles from the periphery of the confinement volume into the center can be realized.

3.3. Simplified Evaluation of the Trajectory Footprint Stochastic Diffusion

Now we apply the analytical expressions obtained above to evaluate the direction of the particle trajectory footprint diffusion. It is reasonable to analyze the expressions

$$D_{r,r}\frac{\partial\Psi^*_{0,m,n}}{\partial r} \tag{3.20}$$

And

$$D_{r,\vartheta}\frac{\partial\Psi^*_{0,m,n}}{r\partial\vartheta} \tag{3.21}$$

Which can be considered (by analogy with well known diffusion processes) as the trajectory footprint fluxes in the radial (r) and angular (ϑ) direction respectively. After the substitution (3.6), (3.7), (3.11) and (3.16) the expressions (3.20) and (3.21) take the forms

$$D_{r,r}\frac{\partial \Psi^*_{0,m,n}}{\partial r} = \left(\frac{V_\parallel}{B}\right)^2 \frac{2\pi R}{V} B_0^2 b_{m',n'}^2 \left(\frac{r}{a}\right)^{2(m'-1)} \frac{1}{2}\left[1-\cos 2(m'\vartheta - n'\varphi + \delta_{m',n'})\right]\times$$
$$\times\left[[m\iota(r^2)-n]r + b_{m,n}\frac{R}{a}\left(\frac{r}{a}\right)^{m-1}\cos(m\vartheta - n\varphi + \delta_{m,n})\right], \tag{3.22}$$

$$D_{r,\vartheta}\frac{\partial \Psi^*_{0,m,n}}{r\partial \vartheta} = \left(\frac{V_\parallel}{B}\right)^2 \frac{4\pi R}{V}(-1)B_0^2 b_{m',n'}^2 b_{m,n}\left(\frac{r}{a}\right)^{2m'+m-3}\frac{R}{2}\sin(m'\vartheta - n'\varphi + \delta_{m',n'})\times \tag{3.23}$$
$$\times\sin(m\vartheta - n\varphi + \delta_{m,n}).$$

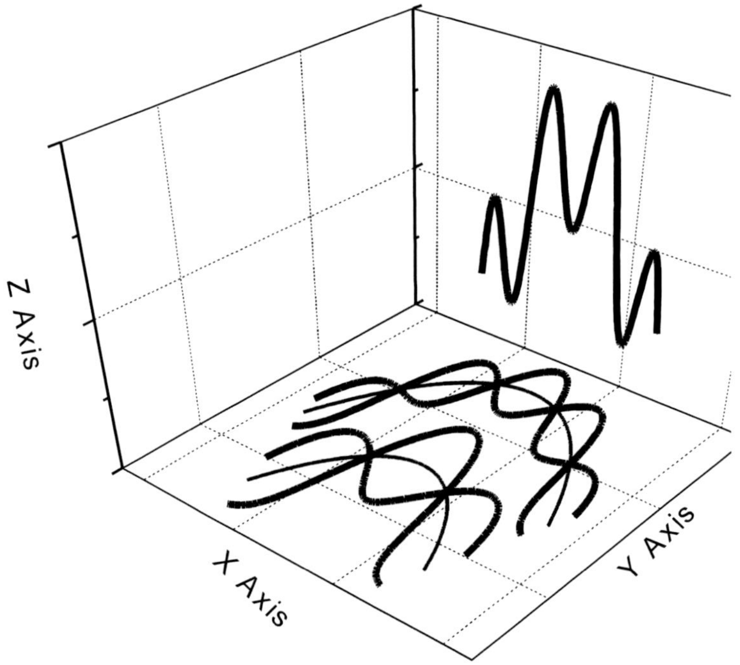

Figure 2. Normalized "trajectory footprint fluxes" (along Z axis) near the (m,n) island separatrix (in the X-Y plane) / distribution in the angular (ϑ) direction.

From these expressions one can find the angle value intervals where the fluxes (3.22) and (3.23) are the largest or smallest ones. On Figure2 geometry factor of normalized flux (3.23) near the separatrix of the (m,n) islands is presented. This distribution is obtained with the following parameters $m=5, m'=9, n=10, n'=10, \delta_{m,n}=\delta_{m',n'}=\pi/2$, $R=390\,cm, a=43\,cm, b_{m,n}=0.003$. One can conclude that in the direction of the separatrix rib ($r_{S,m,n}$, $\vartheta_{S,m,n}$) the flux of the trajectory footprints is the largest. Here one can wait for the capture of the resonant (m,n) particle by the adjacent (next) resonance (m',n').

4. Estafette of Resonance and Helium Ion Removal

In order to perform further investigation and to demonstrate the main physics effects, the numerical integration of the guiding center equation system (2.1) with the use of the main magnetic field in the form (2.2) and the magnetic perturbation field (2.3) is carried out. A fourth order Runge-Kutta method is used for the numerical integration of the guiding center equations.

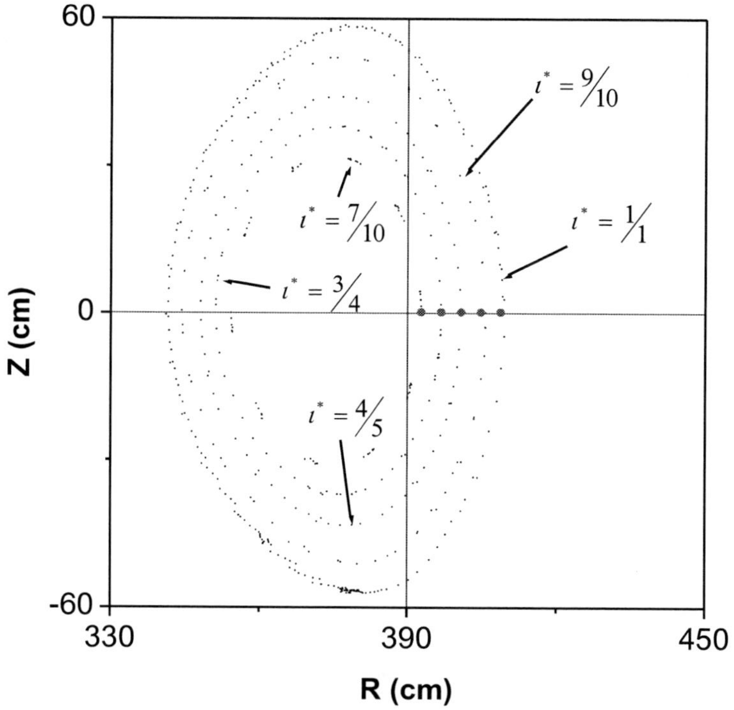

Figure 3. Cross-sections of the drift surfaces of helium ions with the energy W=100 keV, $V_{\parallel}/V$=0.9 with the starting point coordinates r_0=3, 7.1, 11, 15.9, 19 cm, ϑ_0=0 and φ_0=0 without perturbations; time of the observation τ_{observ}=1.22×10^{-3} s for all starting points except r_0 $=19$ cm where τ_{observ} =2.44×10^{-3} s.

For analysis here the configuration for the Large Helical Device (LHD) – the most advanced steady-state helical device which is under successful operation at the National Institute for Fusion Science (Toki, Japan) – is chosen. The possibility of the experimental study of cold alpha-particles on LHD (simulation experiment) has already been discussed [22]

4.1. Test Particle

Helium ions ($_2He^4$) with the energy W=100 keV and starting pitch $V_{II}/V = 0.9$ are taken as the test particles. Starting point coordinates are r_0=3, 7, 11, 15, 19 cm, at ϑ_0=0 and φ_0=0. Here r_0, ϑ_0 and φ_0 are the coordinates of the starting (launching) points of the particles. The starting points in the Figures below are designated with the large crosses. It should be noted that there is a difference between the radial coordinates of the launching points (which are referenced from the circular axis of the torus) and the averaged radii of the drift surfaces because the drift surface axis shifts relative to the position of the circular axis of the torus.

4.2. Drift Surfaces without Perturbations

The so-called Inward Shift configuration is taken where the magnetic axis is shifted toward the torus center and the rotational transform varies from $\iota(0) \approx 0.6$ till $\iota(a_p) \approx 1.0$. Thus rational values of $\iota(r^2/a^2)$ such as 0.75, 0.8, 0.9,1.0 can be used. Among the rational values of the drift rotational transform angle the following set is chosen $\iota(r^2/a^2) \approx$ 0.75, 0.8, 0.9,1.0 (Figure3) and the corresponding magnetic field perturbations can be introduced. Below in subsection 5.1, where the coil system for producing the perturbations is discussed, there is an explanation for why namely these resonances are used here.

4.3. Drift Surfaces with Perturbations

4.3.1. Splitting of Resonant Surfaces

All perturbations act (switched on) simultaneously

Under the set of perturbations: m_p=4, n_p=3; $\varepsilon_{4,3,p}$= 0.003, $\delta_{4,3,p} = \pi/2$; m_p= 5, n_p=4 $\varepsilon_{5,4,p}$=0.007, $\delta_{5,4,p} = \pi/2$; m_p=9, n_p=10; $\varepsilon_{9,10,p}$= 0.03, $\delta_{9,10,p} = 0$; m_p= 5, n_p=5, $\varepsilon_{5,5,p}$=0.00, $\delta_{5,5,p} = 0$, - some families of drift islands in the vertical cross-section of torus appear (Figure 4). For the drift surfaces with the drift rotational transform angle ι^* =3/4 four drift islands take place, for ι^* =4/5 five drift islands appear, the surface with ι^* =9/10 does not split but is corrugated and one island appears instead at the drift surface with ι^* =1/1. It should be noted that the amplitude of the perturbation with m_p=5, n_p=5 is equal to zero. The one island at ι^* =1/1 is the result of the interaction of all magnetic field harmonics.

4.3.2. Stochasticity of Drift Trajectories

Perturbations switched on in the strict sequence

This is the most interesting effect. The perturbations are switched on in a strict sequence as is shown in Figure 5 a. The amplitudes of the perturbations with $m_p = 5$, $n_p = 4$ and $m_p = 9$, $n_p = 10$ are 1.4 times larger than in Figure 4. Another very important fact is that the perturbation with $m_p / n_p = 5/4$ is switched on at the moment of time when the radial deviation of the test particle is the largest. Therefore the test particle starting at the point with $r_0 = 7$ cm, $\vartheta_0 = 0.0$, $\varphi_0 = 0.0$ can move from the initial magnetic surface to the following magnetic surfaces, i.e. from the center of the confinement volume to the periphery. This is seen in Figure 5 b and c. The radial variable of the particle trajectory increases in time and achieves a value at the periphery of the confinement volume (marked by a "cross" in Figure 5 d) where it may be removed by mechanical means. It is possible to mark out the interval of time when the particle is in resonance with the $m_p / n_p = 4/3$ perturbation, then the interval of time when the particle is in resonance with the $m_p / n_p = 5/4$ perturbation and finally the time when it leaves the plasma. Stochasticity of the trajectories and the diffusive character of the trajectory with different but rather close starting points and their reciprocal penetration is shown on Figure6. We would like to emphasize that in Figure5 d the footprints of the full trajectory are gathered together while in Figure 6 (and Figure 3, 4b, 5 b) only the footprints from the "same name" vertical cross-sections are taken into account. To observe the stochastization of the particle trajectory is more convenient to use the vertical cross-sections of the drift surfaces while to evaluate the maximum deviation of the radial variable and the deviation of the angular variable of the particle it is necessary to use the full particle trajectory.

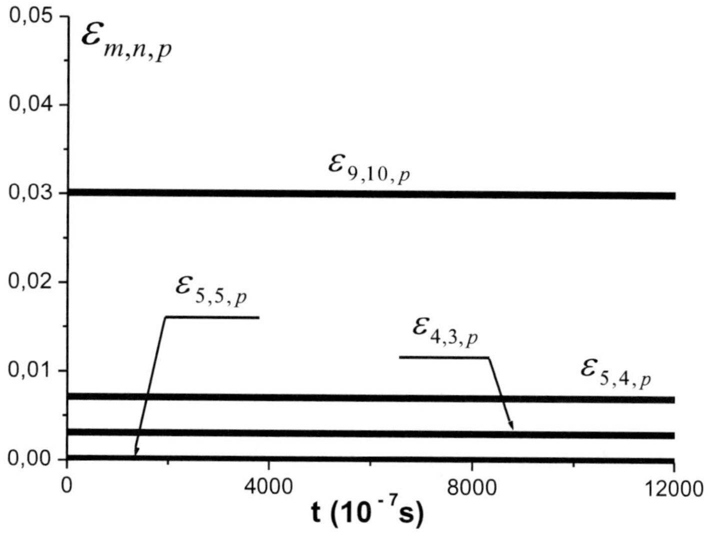

Figure 4. a). Amplitudes of the perturbations in the case when all of them are switched on simultaneously.

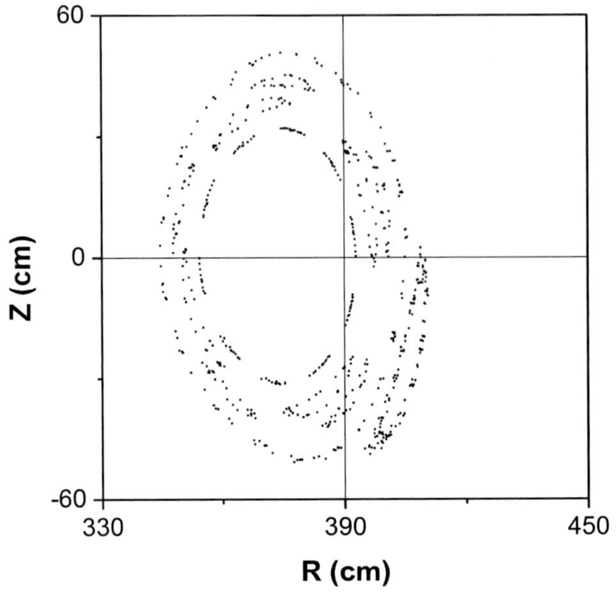

Figure 4. b). Cross-sections of the drift surfaces of the helium ions with the energy W =100 keV, $V_{\|} / V$ =0.9 with the starting point coordinates r_0=3, 7, 11, 15, 19 cm, ϑ_0=0 and φ_0 =0 under the set of perturbations indicated in Section 4.3.2. The time of the observation is τ_{observ}=1.22$\times 10^{-3}$ s.

We have considered here two cases.

In the first case the perturbations are switched on simultaneously but they have the small amplitudes and do not lead to stochastisation of the trajectory. The drift islands are present but they are not overlapped. The deviation of the particle from the initial surface is the island width size.

In the second case the perturbations with the larger amplitudes are switched on consecutively in time. The test particle escapes from the confinement volume following the resonances consecutively, one after another as is expected in accordance with the *estafette* of resonances. The second case is preferable because the "cold" alpha-particle escapes and the drift surfaces of the bulk plasma ions and electrons are not destroyed.

If these perturbations (with larger amplitudes) are switched on at the same moment of time (all of them from the beginning) the particle escapes from the confinement volume earlier but destruction of drift surfaces of other particles, particularly bulk plasma ions takes place.

One practical question is how the switching times of perturbation modes are chosen. Here we follow the particle trajectories and fix the time when the trajectory deviation in the radial direction (variable r) achieves the maximum value under the resonance $\iota^* = n_p / m_p$. Then the perturbation with the following set of "wave" numbers (n'_p, m'_p) satisfying the resonance condition $\iota^* = n'_p / m'_p$ is switched on and this process should continue.

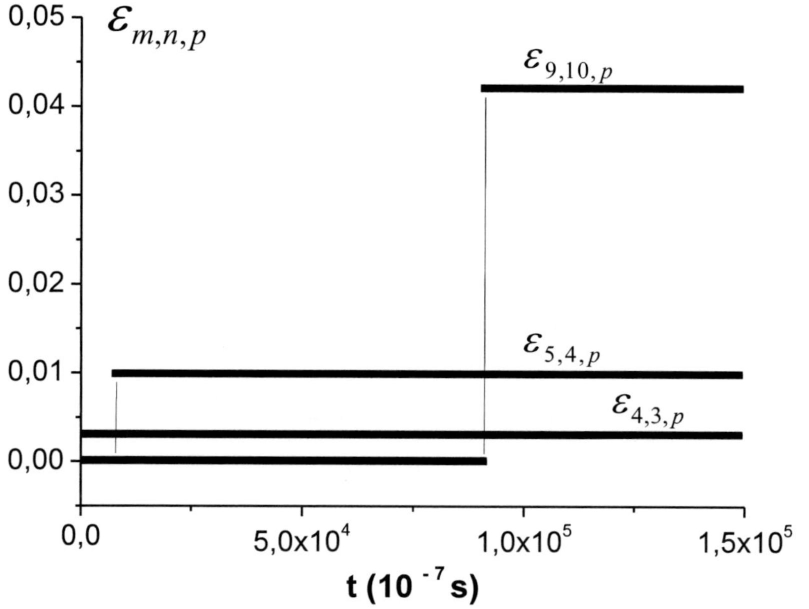

Figure 5. a) Amplitudes of perturbations in dependence on time in the case when they are switched on in a strict sequence.

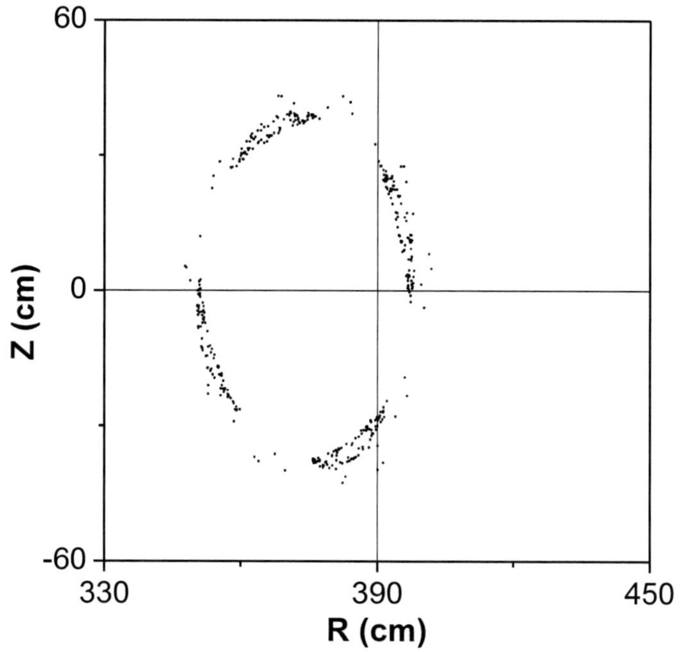

Figure 5. b) Initial stage of the capture of the (m, n) resonant capture by the adjacent (next) (m', n') resonance, the time of observation is τ_{observ}=0.43×10^{-2} s.

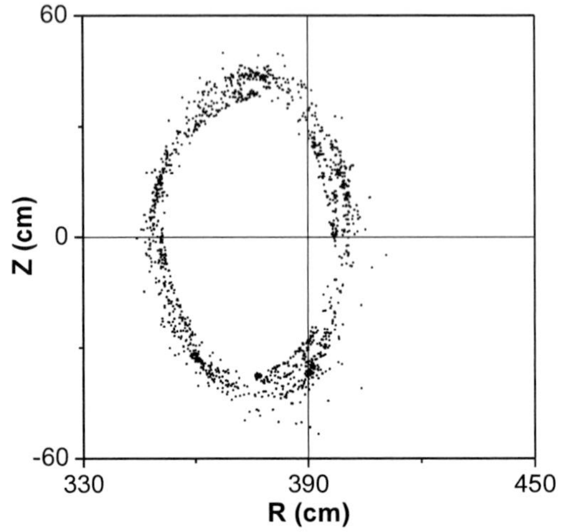

Figure 5. c) Diffusion of the drift surface of the test helium ion with the starting point r_0=7 cm; the time of observation is τ_{observ}=1.5×10^{-2} s.

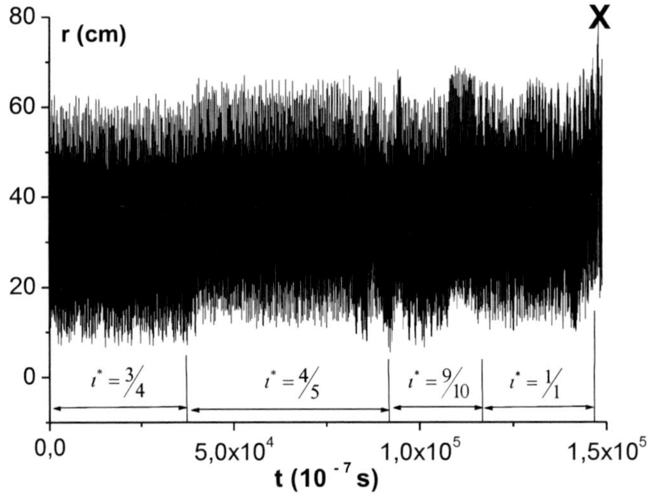

Figure 5. d) Radial variable of the test particle trajectory in dependence on time; the time of observation is τ_{observ}=1.5×10^{-2} s.

4.3.3. Selectivity of Stochastic Transport

One important question is why the bulk plasma ions and electrons do not escape while the helium ion leaves the confinements volume. First of all below we describe the examples of the trajectories of the bulk plasma ion and electrons starting at the same launching point (initial point) as the helium ion. Then we summarize the computation results for all launching points taken here (Figure 7).

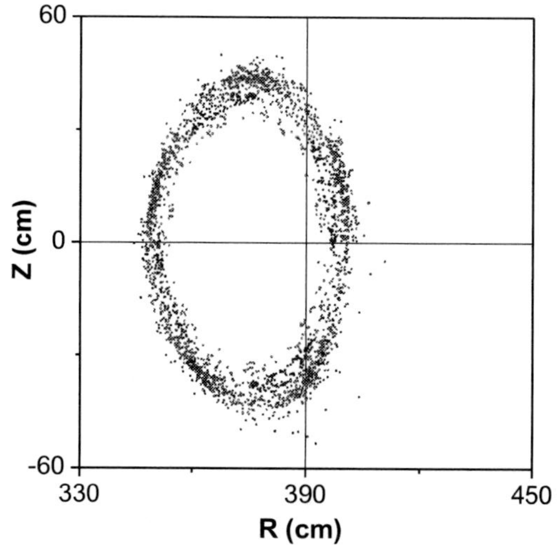

Figure 6. Diffusion of two drift surfaces for the test helium ions with the starting points r_0=7 cm (blue) and r_0=11 cm (red).

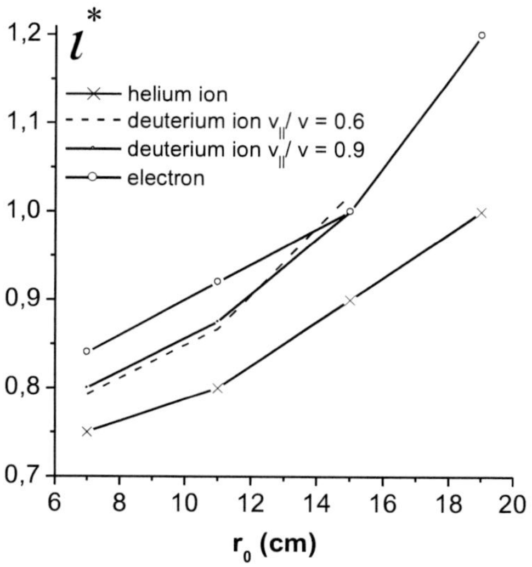

Figure 7. Computational drift transform angle ι^* versus the launching point radial coordinate r_0 (ϑ_0 =0, φ_0=0) for helium ion with $W = 100\,\text{keV}$, $V_{II}\,/\,V$ =0.9, electron and deuterium ion with $W = 7$ keV, $V_{II}\,/\,V$ =0.9 and $V_{II}\,/\,V$ =0.6.

Bulk Plasma Ions Motion

If a deuterium ion with energy W =7 keV (the possible thermal energy of the bulk ion plasma in the device considered) starts at the point where its trajectory drift rotational transform ι^* =4/5 then this particle forms 5 islands in the vertical cross-section instead of 4 islands as in the case of the helium ion with W = 100 keV. The trajectory of the particle with energy W =7 keV is not stochastic, the islands are not destroyed.

Bulk Plasma Electron Motion

Electrons with the energy $W = 7$ keV starting at the same point have the drift rotational transform ι^{*} =0.836 (computational value). Their drift surface have some corrugation but are not stochastic.

The computational drift angle of the bulk plasma electron and ion is different from the computational drift angle of helium ion (Figure7). That is why the bulk particles are not involved in the process of *estafette* of resonances and do not escape from the confinement volume.

4.3.4. Fraction of Helium Particles Involved in Stochastic Transport (Statistics)

Helium ions with energy $W = 100$ keV , starting pitch $V_{II} / V = 0.9$ and with starting points as shown in Figure 8(open circles) come to the periphery of the confinement volume (closed circles). These points are chosen in such a way that they gather in the upper part of the confinement volume. These particles can be collected and removed by mechanical means. The footprints of particles are shown on the background of the drift surfaces from Figure 4 b. The time for escape of the test particles is approximately the same and is equal to 1×10^{-3} s.

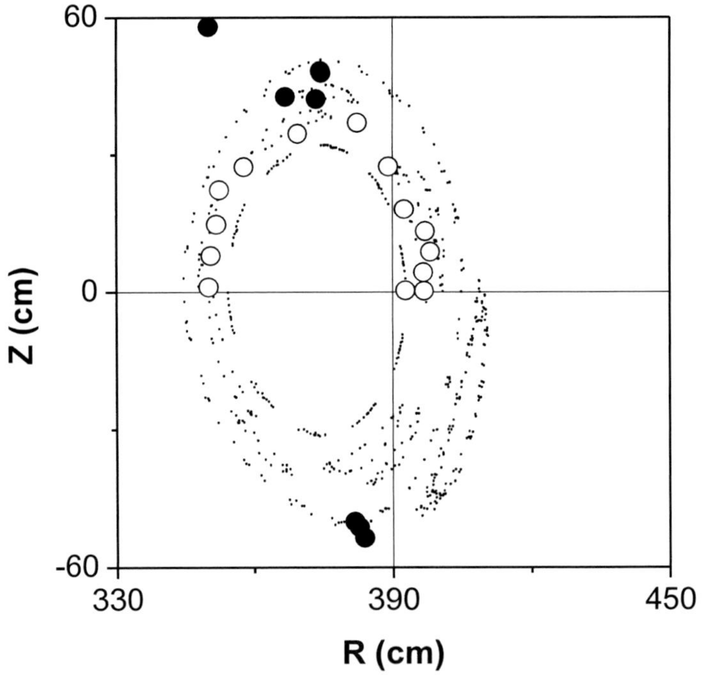

Figure 8. Statistics for the removal of helium ions with W =100 keV, $V_{\parallel} / V$ =0.9 . Starting points are denoted with open circles and the points of maximum distance in the Z direction are denoted with closed circles.

5. Discussion

5.1. Coil System for Producing the Perturbations

There exist some practical ways to produce the perturbations and create the conditions for diverting the plasma with the use of stochasticity, for example [23-29]. For the purpose of the *estafette* of resonances a system like the one described in [23-25] can be used. The number of coils N should be proportional to the product of m_p for the set of perturbing harmonics, more exactly to the least common multiple of the numbers in the denominator of the rational values. Here a set of perturbations is used which corresponds to the following rational values: 3/4, 4/5, 9/10,1/1. Of course, there exist other resonances, such as 2/3, 5/6,6/7, 7/8, 8/9 etc. However, only several main resonances, namely 3/4, 4/5, 9/10 1/1 are used in this study. The reason is that the product of the numbers which are in the denominator of the rational values should determine the number of the current ring coils which can produce the magnetic field perturbations with the corresponding set of "wave" numbers.

It is desirable that the number of the coils, N, would be not too large.

It is possible to use a more internal resonance such as drift surface with $\iota^* = 2/3$ and magnetic perturbation with $m_p / n_p = 3/2$. In this case the family of 3 islands appear in the meridional cross-section. The dimensions of other islands and resonant surfaces change too.

5.2. Effect of Other Physics Mechanisms

The electric field does not affect the resonance condition for high energy particle ("cold" alpha-particles) strongly. Stochasticity takes place if an electric field with $e\Phi / W = \pm 0.1$ is present.

The effect of scattering [30] on the resonance studied in [31] has shown that there exist the conditions when the resonances are not broken.

5.3. Other Possibilities to Use the Described Mechanism

The physical mechanism described here can be applied to solve the "reverse" task- the transport of particles injected from outside the plasma into the center of the confinement volume .In such a way it is possible to inject fueling particles if it is necessary.

5.4. About the Onset of the Chaotic Motion in Deterministic Hamiltonian Systems

As it is known (see for example [10, 11, 16, 33-35]) the chaotic motion can be the property of many dimensional dynamic systems (with many degrees of freedom) and "low number"-dimensional dynamics system (with a small number degrees of freedom) also. The trajectory of the particle motion belongs to "low number "dimensional systems. Such a system also can experience stochastisation . This is a very important issue. The particle trajectory is described by the equation of the non linear oscillator. In the case of the *estafette* of resonances guiding center equations can be reduced to the non linear oscillator under quasi periodic perturbation. Analysis of the oscillator equation in phase space is a commonly known tool of study. However practical tasks and many applications need the analysis in real geometry space. Overlapping of resonances of particle trajectories can lead to stochasticity in the real space. It can be seen on stellarators. That is why the description of these effects should be done in real space coordinates. Such analytical description is given in this paper in section 3 and numerical demonstration is presented in Section 4.

5.5. Possible Scenario of the *Estafette* Resonances Application

Here we assume that the helium ions with certain energy and the parallel velocity are at a certain location in the center of the confinement volume and we demonstrate one way to remove them from the center of the confinement volume to the outside with no deterioration of the plasma ion confinement. Another possible scenario can be the following: if the ions which should be removed are observed at a certain place and their energy and parallel velocity are determined then their drift rotational transform angle should be calculated, the set of mode numbers of the externally applied magnetic field perturbations are chosen and after the switching on the corresponding perturbations the particles can be removed.

5.6. Generalization of the stochastic diffusion coefficients in magnetic (Boozer) coordinates for the wide class configuration including plasma equilibrium configurations is carried out in recent papers [36,37].

CONCLUSIONS

The principal possibility of the new physics mechanism - *estafette* of resonances is shown here.

1. *Estafette* of drift resonances can be used for the removal of a test particle, particularly helium ash, when the perturbing magnetic fields are externally applied. For this purpose in the case of particle trajectories with the drift rotational transform $\iota^* = n_p / m_p$ the perturbing magnetic field with wave numbers m_p, n_p leads to drift island formation. If the adjacent drift islands overlap the particle trajectories become stochastic. A particle passes through a set of adjacent resonances and can leave the confinement volume.

2. This mechanism to remove the ash from the confinement volume or to control the particle motion inside the confinement volume is selective relative to the energy. The bulk plasma ions do not experience the stochastisation of their trajectories and do not escape the plasma.

3. The efficiency of this approach is evaluated (partly) from the statistics. For each case a concrete scenario as described above and applied to this mechanism should be developed

4. Perturbing magnetic fields can be produced by a system of coils with specific currents. The system of coils is not so complicated and can be similar to the coils of the local island divertor.

5. One way more to apply this mechanism-estafette of drift resonances - is to consider the MHD perturbations in plasma and the loss of the energetic particles due to the consecutive onset of the resonance perturbations.

ACKNOWLEDGMENTS

This work is partly supported by the Science and Technology Center in Ukraine in the framework of the Project N 2313.

REFERENCES

[1] Shishkin, A.A. Estafette of drift resonance, stochasticity and control of particle in a toroidal magnetic trap, *Nuclear Fusion* v.42 (2002), pages 344-353

[2] Shishkin, A.A. and Zolotukhin, A. V. Drift resonance of impurity ions in a torodal magnetic trap with rotational transform, *Physics of Plasmas* v.7 (2000), pages 5023-5032.

[3] Motojima, O.and Shishkin, A.A.. Drift island motion in helical plasma and its use for ash removal and high energy ion injection, *Plasma Physics and Controlled Fusion* v. 41(1999), pages 227-242.

[4] Shishkin, A.A. and Motojima, O. Coil System for the Resonant Pumping In/Out of Particles in a Toroidal Magnetic Traps. *Journal of Plasma and Fusion Research SERIES*, v.3 (2002), pages 509-512.

[5] Mynick, H.E. Transport of energetic ions by low- n magnetic perturbations. *Phys.Fluids B* 5 (1993) 2461 and Stochastic transport of MeV ions by low- n magnetic perturbations. *Phys.Fluids B* 5 (1993) 1471.

[6] Mynick, H.E., Pomphrey, N. Frequency Sweeping: A New Technique for Energy-Selective Transport ,*Nuclear Fusion* 34 (1994) 1277.

[7] Chirikov, B.V. A universal instability of many-dimensional oscillator system. *Phys. Rep.* 52 (1979) 263.

[8] Bogolyubov, N.N. and Mitropol'skiy Yu.A. Asymptotic Methods in the Theory of Nonlinear Oscillations, Gordon and Breach, New York, 1962 (Russian orginal, Gos.izdat *Phys.-Mat. Lit*, Moscow 1963).

[9] Mitropol'skiy Yu.A. *Averaged Method in Nonlinear Mechanics*, Naukova Dumka, Kiev 1971, 440 pages (in Russian).

[10] Rosenbluth, M.N., Sagdeev, R.Z., Taylor, J.B., Zaslavski, G.M. Destruction of magnetic surfaces by magnetic field irregularities, *Nuclear Fusion* 6 (1966) 297.

[11] Filonenko, N. N., Sagdeev, R. Z., Zaslavski, G.M. Destruction of magnetic surfaces by magnetic field irregularities II, *Nuclear Fusion* 7 (1967) 253.

[12] Stix T.H. Plasma transport across a braided magnetic field. *Nuclear Fusion* 18 (1978) 353.

[13] Mendonca J.T. Diffusion of magnetic field lines in a toroidal geometry, *Phys. Fluids B* 3 (1991) 87.
[14] Hatori T., Irie, H., Abe, Y., Urata K. Chaotic Magnetic Field Line in Toroidal Plasmas., Progress Theor. *Phys. Suppl.* 1989, No.89, p. 83 and Institute of Plasma Physics, Research Report IPPJ-911, May 1989, Nagoya , Japan.
[15] Rechester, A.B., Rosenbluth, M.N. Electron heat transport in a tokamak with destroyed magnetic surfaces. *Phys.Rev.Lett.* 40 (1978) 38.
[16] Grishchenko, A.D., Vavriv, D. M. Dynamics of a pendulum with a quasiperiodic perturbation. *Tech. Phys.* 42 (1997) 1115.
[17] Morozov, A.I., Solov'ev, L.S. Motion of Charged Particles in Elelctro-magnetic Fields, Reviews Plasma Physics (edit. by Acad. M.L.Leontovich) vol.1, Consultants Bureau, New York, 1996, p.201-297.
[18] Iiyoshi, A., Fujiwara, M., Motojima, O., Ohyabu, N. Yamazaki, K., Design study for the Large Helical Device. *Fusion Technology* 17 (1990) 169.
[19] Shishkin, A.A., Watanabe, K.N., Yamazaki, K., Motojima, O., Grekov,D.L., Smirnova, M.S., Zolotukhin, A.V., "Some Features of Particle Orbit Behavoir in LHD Configurations" National Institute for Fusion Science, Research Report NIFS Series, NIFS – 211, March 1993, Nagoya, Japan, 44 pages.
[20] Popryadukhin A.P. Perturbations of the particle motion in a stellarator. *Plasma Physics*, 8
(1966) 169.
[21] Solov'ev, L.S. and Shafranov, V.D. Plasma Confinement in Closed Magnetic Sysytems (Reviews of Plasma Physics vol.5) ed. M.A.Leontovich (New York: Consultants Bureau)
pp 1-247.
[22] Sato, K.N., Murakami, S., Nakajima, N., Itoh, K. Possibility of Simulation Experiments for Fast Particles in the Large Helical Device (LHD), *Nuclear Fusion* 35 (1995) 1563.
[23] Komori, A. Ohyabu, N., Watanabe, T. et al., Plasma Physics and Controlled Nuclear Fusion Research, 1994, Seville, Local island divertor concept for LHD, *Proc. 15th Intern. Conference, Seville*,1994, IAEA, Vienna, 1995, vol.2, p.773.
[24] Ohyabu, N., Komori, A., Akaishi, K. et al., Innovative divertor concepts for LHD, *Journal of Nucl. Materials* 220-222 (1995) 298.
[25] Motojima, O. , Ohyabu, N., Komori, A. et al., Divertor experiment in large helical device, *Plasma Physics and Controlled Fusion* 38 (1996) A77.
[26] Shabrov, N.V., Khvesjuk, V.I., Lyakhov, A.N. Selective Ions Pumping from an Axisymmetric Mirror Trap, Proc. International Conference on Open Plasma Confinement Systems for Fusion, ed. A.Kanantsev, Budker Institute of Nuclear Physics, Siberian Branch, Russian Academy of Sciences, 1993, World Scientific, Singapore 1994, pp.245 – 260.
[27] Finken, K.H., Abdullaev, S.S., Kaleck, A., Wolf G.H. Operating space of Dynamic Ergodic Divertor for TEXTOR –94, *Nuclear Fusion* 39 (1999) 637.
[28] Kobayashi, M., Tuda, T., Tashiro, K. et al., Interaction of Externally Applied Rotating Helical Field with Tokamak Plasma, *Nuclear Fusion* (to be published).
[29] Kikuchi, Yu., Tashiro, K., Kobayashi, M. et al., Improvement of Local Helical Coil Configuration for Dynamic Ergodic Divertor. *10th International Toki Conference on*

Plasma Physics and Controlled Nuclear Fusion (ITC-10) "Physics and Technology for Steady State Plasmas", January 18-21, 2000, Toki-city, Japan, PI-22, Abstracts, p.74.

[30] Sivukhin, D.V. Coulomb Collisions in a Fully Ionized Plasma, *Reviews Plasma Physics* (edit. by Acad. M.L.Leontovich) vol.4, Consultants Bureau, New York, 1996, p.93-241.

[31] Shishkin, A.A., Motojima, O., Polunovskij, E.I. The Motion of Drift Island in Toroidal Magnetic Trap under Coulomb Scattering, *Tech. Phys. Lett.* 25 (1999) 725.

[32] Shishkin, A.A., Sidorenko, I.N., Wobig, H. Dynamical Control of Drift Resonances in HELIAS Configuration, 6th Ukrainian Conference and School on Plasma Physics and Controlled Fusion, Alushta , Ukraine, September 14-20, 1998, Report A –18; Problems of Atomic Science and Technology, SERIES: Plasma Physics, ISSUES 1(1), 2(2), National Sciende Center "Kharkov Institute of Physics& Technology, Kharkov, 1999.

[33] Klafter, J., Shlesinger, M.F. and Zumofen, G. Beyond Brownian Motion, *Physics Today* 49 (2) (1996) 33.

[34] Lichtenberg, A.J. and Liberman, A. Regular and Stochastic Motion, Springer-Verlag, New York, 1983.

[35] Zaslavsky, G.M. *Chaos in dynamical systems*, Harwood, Chur, 1985

[36] Nickolai Azarenkov, Oleg Shyshkin, Ralf Schneider, Horst Wobig, Yurii L Igitkhanov. Stochasticity of heavy ion trajectories in drift optimized plasma configurations *Journal of Kharkiv University* No.574, 2002, physical series "Nuclei, Particles, Fields" issue 4 /20/, pages 27-43.

[37] Oleg Shyshkin, Ralf Schneider, Yurii Igitkhanov, Stochastic Diffusion Tensor and Tungsten Ions Fluxes In Drift Optimised Plasma Configuration, *14th International Stellarator Workshop* (Greifswald, Germany, September 22 – 26, 2003).

In: Perspectives in Mathematical Physics
Editor: Gregory D. Monteggo
ISBN: 978-1-61122-984-4

Magnetic Reconnection in the Earth's Magnetotail: Reconstruction Method and Data Analysis

T. Penz[1,2,*] *V. V. Ivanova* [3], *V. S. Semenov*[3], *R. Nakamura* [1], *I. B. Ivanov* [4,5], *H. K. Biernat* [1,2], *M. F. Heyn* [4], *V. A. Sergeev* [3], *and I. V. Kubyshkin* [3]

[1] Space Research Institute, Austrian Academy of Sciences, Schmiedlstr. 6, A–8042 Graz, Austria

[2] Institute of Physics, University of Graz, Universitätsplatz 5, A-8010 Graz, Austria

[3] Institute of Physics, State University St. Petersburg, 198504 Russia

[4] Institute for Theoretical Physics, Technical University Graz, Petersgasse 16, A-8010 Graz, Austria

[5] Petersburg Nuclear Physics Institute, Gatchina, 188300 Russia

Abstract

Magnetic reconnection is a topological restructuring of magnetic field lines, leading to a conversion of magnetic energy into kinetic energy and a heating of the plasma. It takes place in regions with strong magnetic gradients: the Earth's magnetopause and the plasma sheet in the magnetotail, during solar and stellar flare events, and in laboratory plasmas on the Earth. It is also involved in astrophysical phenomena like the generation of relativistic jets and processes in accretion disks. The Earth's magnetosphere is the only region where signatures of magnetic reconnection can be measured in-situ by satellites. Based on the method developed by Semenov et al. [1], we solve the direct problem of magnetic reconnection and calculate the magnetic field and plasma flow disturbances in a compressible plasma caused by transient reconnection. In order to get the reconnection rate from measured data, we invert the problem. Since the inverse problem is ill–posed, it is treated with the method of regularization. Application of this method to Cluster measurements from September 8th, 2002, where a series of Earth-ward propagating 1–minute scale magnetic field and plasma flow variations were observed outside of the plasma sheet, showed good agreement for the z-component of the reconstructed magnetic field. The reconnection rate is about 1 mV/m and the reconnection region is located at about 24–25 Earth radii in the magnetotail.

Keywords: Magnetic reconnection, Magnetotail, Cluster, Magnetohydrodynamics

*E-mail address: thomas.penz@oeaw.ac.at

1. Introduction

Magnetic reconnection is a fundamental plasma process allowing the conversion of magnetic energy into kinetic and thermal energy of the plasma. The occurrence of thin current layers associated with a plasma flow and the presence of magnetic fields inevitably leads to the process of magnetic reconnection. In fact, magnetic reconnection is caused by a change in the plasma resistivity leading to the release of stored magnetic energy.

On scales larger than the characteristic length of the plasma (e.g., the gyroradius), the motion of the plasma is coupled with the magnetic field. This is called the frozen–in condition, which implies that plasma and magnetic fields from distinct regions do not intermix and stay separated. In the upper panel of Fig. 1, this situation is displayed for the simple configuration of two antiparallel magnetic fields, which are kept apart by a thin current layer. If magnetic flux is added to the current sheet from both sides, free energy is accumulated in the current sheet. After some time, the gradients across the current sheet become too strong, resulting in a violation of the frozen–in condition. If such a violation takes place, reconnection is initiated.

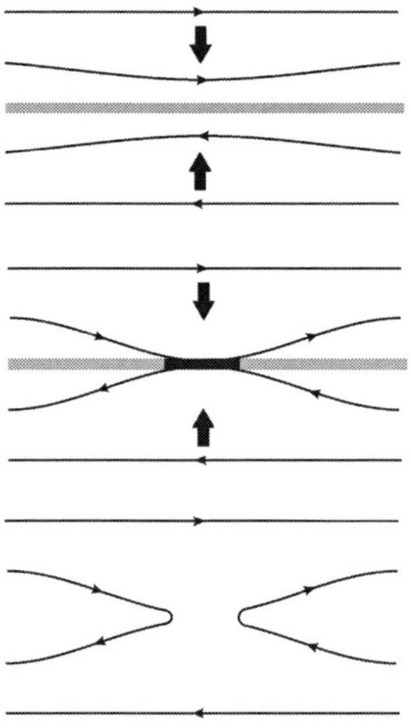

Figure 1. Different stages during the process of magnetic reconnection. The upper panel shows the initial configuration of antiparallel magnetic fields separated by a thin current sheet (dashed area). If the magnetic fields are convected towards each other, strong gradients occur, leading to the formation of a diffusion region (black area), where the frozen–in condition is not valid anymore (middle panel). In this region, the magnetic field lines can be broken and reconnected (lower panel). This leads to a connection between plasma regions which were previously separated.

Because of the highly non–uniform structure of the current sheet, it is very likely that a local breakdown of the frozen–in condition takes place, leading to the formation of the so–called diffusion region. Early models of magnetic field annihilation as proposed by

Sweet [2] and Parker [3] assumed a diffusion region with a size comparable to the whole current sheet. It was quickly realized that this process is much too slow to account for rapid energy release expected in many astrophysical processes. A way to avoid this problem was proposed by Petschek in 1964 [4]. In contrast to the Sweet–Parker model, where homogeneous resistivity is assumed, Petschek proposed a localized enhancement of the resistivity in a highly conducting fluid. Thus, the diffusion region, where the assumption of an ideal conducting plasma does not hold, becomes much smaller than the scale size of the system (the black area in the middle panel of Fig. 1). Petschek showed that the diffusion region is the source for large amplitude MHD waves. In this model, most of the inflowing plasma crosses the shock waves and not the diffusion region anymore. From this point of view, the shock waves transport information into a wide region of ideal plasma, allowing for much higher reconnection rates than in the Sweet–Parker model. It was also shown by simulations using resistive MHD equations that the Sweet–Parker configuration appears for a spatially constant plasma resistivity, while Petschek–type structures arise for a strongly localized plasma resistivity [5, 6]. This insight is also supported by numerical simulations [7]. The newly formed region bounded by the shock waves is called outflow region, where the inflowing plasma is heated and accelerated, leading to the formation of bi–directional jets [8] and fast plasma flows, which are important features indicating the occurrence of magnetic reconnection. Also the connection of the magnetic field lines takes places over the outflow region (lower panel of Fig. 1).

In the last century, many sites were discovered where magnetic reconnection plays a crucial role [9]. It is observed in laboratory fusion machines like tokamaks [10, 11]. In order to investigate the fundamental physics involved in magnetic reconnection in laboratory settings, different devices like the Magnetic Reconnection Experiment (MRX) device were constructed [12]. It revealed the appearance of different shapes of the diffusion region as well as the importance of the shear angle between the magnetic field lines and the influence of boundary effects.

Moving away from the Earth, other plasma environments suitable for reconnection can be found. The first environment where magnetic reconnection was expected is the Sun. The earliest models [13, 2] have aimed at explaining the large amount of energy released in solar flares. Reconnection on the Sun is inferred from different observations, including high speed flows of X–ray and EUV features, and energetic particle events in hard X–ray images [14, 15], which have been observed by several spacecrafts like Yohkoh, SOHO, and TRACE. The application of Petschek–like reconnection to solar flares gives typical reconnection times consistent with observations [16]. Reconnection has also been proposed as a mechanism for the heating of solar and stellar coronae to extremely high temperatures of more than 10^6 K [17]. The importance of reconnection in plasma environments outside of our solar system is still under debate. The two astrophysical topics to which reconnection theory has been extensively applied are stellar flares [18] and accretion disks [19]. Reconnection theory also enters into models of protostellar collapse, extragalactic jets, galactic magnetic fields, and even galactic clusters [9].

However, the only place in space were magnetic reconnection can be observed in–situ is the terrestrial magnetosphere. At the dayside magnetopause, energy and plasma from the solar wind can enter the terrestrial magnetosphere because of the reconnection of the interplanetary magnetic field (IMF) and the Earth's magnetic field [20]. Of particular interest is the transient feature of magnetic reconnection, leading to the formation of so–called flux transfer events (FTEs), which are flux tubes connecting the solar wind plasma with the magnetospheric plasma. The first in-situ measurements of flux transfer events caused by transient reconnection were reported by Russell and Elphic in 1978 [21] for the dayside magnetopause. They defined FTEs as signatures with a bipolar variation of the magnetic field component perpendicular to the current sheet, a deflection of the tangential component, and an increase in the magnetic field strength. In the same year, Haerendel et al. [22] found indications for reconnection processes also at the high–latitude dayside magnetopause. In the following years, a collection of studies supported the interpretation that FTEs are burst of magnetic reconnection at the magnetopause appearing most frequently during intervals of southward directed IMF [23, 24]. However, newer observations with the Cluster satellites show that for a northward–orientated IMF, reconnection can appear tailward of the cusp [25]. There are also strong indications that reconnection at the dayside magnetopause can be continuous over an extended period of time [26].

Magnetic flux reconnected at the dayside is convected into the tail lobes, where reconnection is initiated in a region between 15 and 30 R_e in the magnetotail [27, 28]. It was proposed by Sergeev et al. [29] that FTE–like structures appear also in the Earth's magnetotail. They performed a qualitative comparison between measurements of ISEE–1 and model predictions from a transient reconnection model, which showed a general agreement. In later work, these events were referred to as nightside flux transfer events (NFTEs) [30]. These are short–term events in the substorm–time plasma sheet, which can be described by impulsive variations of the reconnection rate in models of transient reconnection. They are characterized, similar to FTEs, as a bipolar variation of the B_z–component and a deflection of the x–component of the magnetic field. Additionally, the velocity shows an upward flow of plasma in the beginning, followed by a strong flow directed downward to the current sheet. Such structures noticed in the tail plasma sheet are often referred to as individual bursts of BBF, as transient plasma sheet expansions, as plasmoids or flux ropes, as well as NFTEs [30, 31, 32]. The appearance of reconnection structures in the magnetotail was also confirmed by several other satellite missions, like Geotail [33] or Cluster [34, 35].

The aim of this paper is to show recent evidence for magnetic reconnection from Cluster measurements and introduce a model for the reconstruction of the reconnection rate from satellite measurements. In Section 2, multi–spacecraft measurements of the Cluster mission are discussed with emphasize to the implications for magnetic reconnection. The theoretical model used to analyze NFTEs in the terrestrial magnetotail is described in Section 3. Section 4 is devoted to the reconstruction of the reconnection rate from Cluster measurements. Also the site of reconnection with respect to the satellite is estimated.

2. Observational Features of Magnetotail Reconnection

Whereas the dayside magnetosphere is compressed by the solar wind, the nightside magnetosphere is stretched out into a long magnetotail. When the interplanetary magnetic field (IMF) is southward, the merged field lines at the dayside are transported towards the nightside, reconnected in the distant tail and then transported back towards the dayside, creating inward and Earthward flows in the central plasma sheet. The flux transport rates at the dayside and at the nightside are, however, balanced only in an average sense and the unbalanced

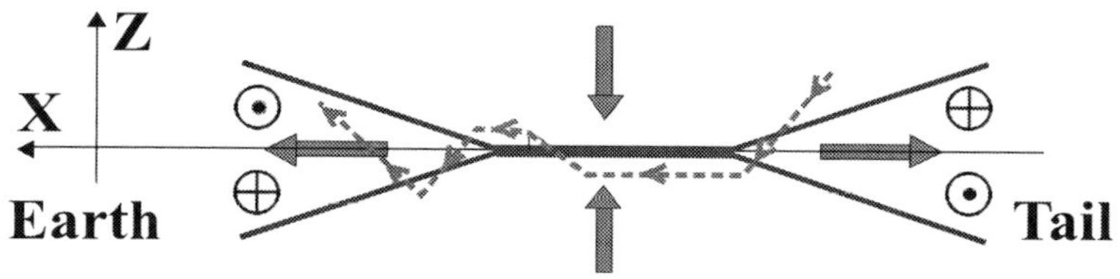

Figure 2. Sketch of Cluster observations of Hall magnetic fields and current sheet structure around the reconnection region; the red dashed line indicates the trajectory of Cluster (after Runov et al. [34]).

transport is the ultimate cause of a substorm when magnetic reconnection takes place closer to the Earth, at radial distance of 15 - 30 R_E in the tail. The processes associated with this near-Earth reconnection determine different scales of magnetotail dynamics.

Cluster traversed the magnetotail covering regions Earthward of 19 R_E during the past summer seasons since July 2001. The tetrahedron scale was between 250 km and 4000 km, which was changed every year, so that characteristics at different scales could be identified. The four spacecraft observations enable us for the first time to differentiate spatial from temporal disturbances. In this section, we highlight several Cluster observations when Cluster was located near the X-line and more remote observations showing the effects of the processes in the ion diffusion region.

2.1. Observation near the X-line

The four spacecraft observations enable us to obtain spatial gradients of different parameters continuously. For example, the current density obtained from the gradient in the magnetic field is an essential parameter for magnetotail diagnostics. Using the gradient information obtained from the four Cluster spacecraft during current sheet crossings, current sheet structures in the vicinity of X-line were reconstructed by Runov et al. [34] as illustrated in Fig. 2. A thin current sheet with a half-thickness of about one ion gyro radius was found for the crossing closest to the X-line, whereas the outer crossings showed bifurcated current sheet profiles. Changes in the curvature of the field for the different current sheet crossings illustrated in the figure were consistent with a X-line motion from Earthward and tailward of the spacecraft. Furthermore, consistent features of the field disturbance associated with Hall-current at both side of X-line were identified, confirming that the spacecraft traversed the ion diffusion region.

The importance of multi-scale processes from heavy ions to electron kinetics in the reconnection processes have been identified by several Cluster observations. During thin current sheet intervals associated with crossings of the X-line during storm-time substorms, the O^+ pressure and density were observed to be dominated compared to those of H^+ [36]. In such current sheets, the O^+ were observed to execute Speiser-type serpentine orbits across the tail and were found to carry about 5-10% of the cross-tail current [36]. The O^+ in the reconnection region was suggested to experience a ballistic acceleration [37] based on the observation of a large amplitude bipolar electric field. These large-amplitude (up to 50 mV/m) solitary waves, identified as electron holes, have been observed during several plasma sheet encounters that have been identified as the passage of a magnetotail reconnection X-line [38]. The electron holes were seen near the outer edge of the plasma sheet, within and on the edge of a density cavity, at distances in the order of a few ion inertial lengths from the center of the current sheet. Based on a detailed comparison with simulations, Cattell et al.[38] suggested that the observed nonlinear wave modes, electron holes, may play an important role in reconnection by scattering and energizing electrons.

2.2. Observations Away from the Reconnection Site

Whereas in situ observations of the ion diffusion region depend on the rare chance that the spacecraft being located at the right place at the right time, different effects of reconnection can be detected remotely from the reconnection site and can still contain useful information on temporal and spatial characteristics of reconnection. For example, the limited area of the Hall currents flowing in the ion diffusion region indicates that the current has to be closed including regions outside of the ion diffusion region. At the lobe side, the closure of the Hall current takes place via cold electron flowing into the ion diffusion region. At the outflow region, on the other hand, the accelerated electrons can carry the current into the ion diffusion region. Cluster observed such field-aligned electrons related to the Hall current system consistent with previous observations, but also obtained fine structures indicating multiple temporal or spatial properties of reconnection [39, 40, 41]. Fig. 3 illustrates multi-point observations by Cluster in the northern hemisphere and by Geotail in the southern hemisphere detecting these field aligned currents possibly connected to the ion diffusion region from Nakamura et al. [39]. The four Cluster satellites also allowed to determine the scale size of these field aligned currents, suggesting that the scale size of the downward current was at maximum comparable to the ion inertia length so that it plausibly connects to the near-Earth X-line and is driven by Hall effects in the reconnection region. This interhemispheric observation supported the theoretical prediction [42] that the downward current region is thin because on the lobe side, the ions may travel a substantial fraction of the ion inertia length until their motion separates from that of the electrons.

Using multi-composition plasma observation by Cluster, slow-mode shocks connected to the ion diffusion region have been analyzed by Eriksson et al. [43] taking into account also the contribution from oxygen ions during a substorm X-line event when Cluster observed fast tailward and Earthward flows. The successful joint Walén and slow shock analyses on the tailward flows within the plasma sheet presented further evidence in favor of Petschek-type reconnection at distances $X_{GSM} > -19\ R_E$ of the near-Earth magnetotail.

Cluster succeeded to obtain detailed characteristics of Earthward propagating southward then northward magnetic field disturbances related to plasmoids/flux ropes [44, 45], travelling compression regions [46, 47], and nightside flux transfer events [48, 49, 35]. Multi-point analysis by Cluster were used to measure the current density and check the force-free model [46] and energetic particle boundaries [45] to show the structures of the plasmoid/flux rope. Yet, since the plasma flows jetting towards the Earth are significantly influenced by the strong dipolar field and pressure gradient, it still remains unknown to what extent these structures can be treated as motions of a stable structure in the analysis. As described in the later sections, similar magnetic features were rather interpreted as transient profiles of a remote X-line due to its change in the reconnection rate and used to

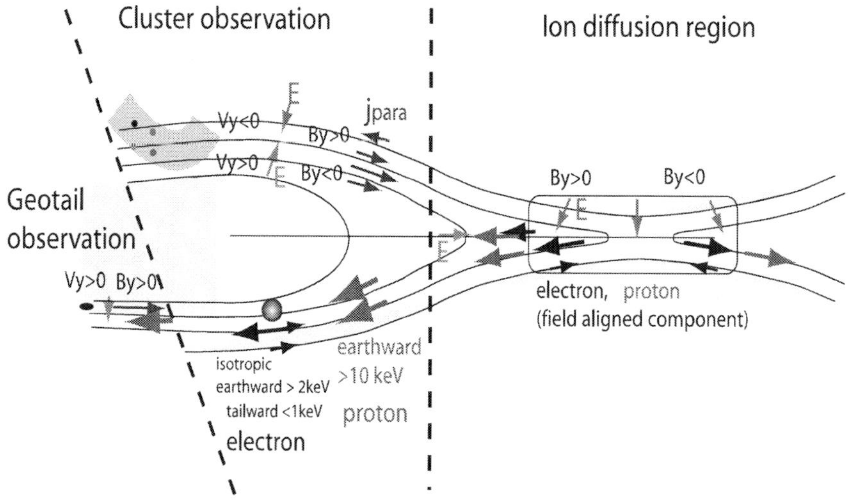

Figure 3. Summary of observations by Cluster and Geotail during a transient entry into the plasma sheetd during a substorm event and illustration of the possible relationship to the reconnection region. For Cluster observations field signatures are showing in the northern hemisphere, while particle signatures are illustrated in the southern hemisphere; adapted from [39].

determine the location of the X-line [48, 49, 35]. Observationally, the determination of the field topology would be a key to differentiate whether these structure are coming from a single X-line or are signatures of multiple X-lines.

In the plasma sheet, bursty bulk flows (BBF) are the most clear signatures of the consequence of reconnection and have been intensively studied using Cluster spacecraft measurements based on multi-point data analysis. A statistical analysis was performed by Nakamura et al. [50] to estimate the typical scale size of BBF. Different types of magnetic field disturbances accompanied by the BBF were studied such as dipolarization [51], low frequency wave activity [52], and turbulence [53]. Using Cluster, together with the other spacecraft and ground-based observations, BBF-associated field-aligned and ionospheric current system have been obtained [54, 55].

3. The Petschek–Type Model of Magnetic Reconnection

In the following, we present a model to reconstruct the reconnection rate developed by Semenov et al. [67] for an incompressible plasma extended to a compressible one. First we show the solution of the direct problem, giving the behavior of NFTEs for a prescribed reconnection rate. To reconstruct the reconnection rate, we solve the inverse problem. This enables us to achieve the reconnection rate from magnetic field measurements.

3.1. Basic Configuration

In order to describe the temporal evolution of FTEs, time–dependent Petschek–type models of reconnection were developed [1, 56, 57, 58, 59]. For our purpose, we consider the structure of the Earth's magnetotail as oppositely directed magnetic fields, which are separated by a current sheet, the so–called plasma sheet. In terms of ideal MHD, this current sheet can be described as an infinitely thin tangential discontinuity. The normal components of the magnetic field and plasma velocity are zero ($B_n = 0$, $v_n = 0$) across a tangential discontinuity, while the density and the tangential components of the magnetic field and velocity can change arbitrarily, subject only to the requirement that the total pressure, which is the sum of the thermal and the magnetic pressure, $p + B^2/(8\pi)$, stays constant [60]. Due to these requirements, there is no mass flow and no magnetic connection across a tangential discontinuity. As a result, there will be no electric field component along the discontinuity ($E_t = 0$).

Due to the local enhancement of the plasma resistivity, a tangential electric field is generated, causing a normal component of the magnetic field. Thus, the conservation laws for a tangential discontinuity are violated, and the surface gets non–linearly unstable, decaying into a system of MHD wave modes according to the general Riemann problem [61, 60]. The most general scheme of a decay looks like $S^+(R^+)AS^-(R^-)C(T)S^-(R^-)AS^+(R^+)$, where S^+, S^- are fast and slow shocks, R^+, R^- are fast and slow rarefaction waves, and A, C, T are Alfvén, contact, and tangential discontinuities, respectively. If we assume that the total pressure is conserved in lowest order, no strong fast shocks are present in the system. For symmetric conditions, no contact discontinuity and no rarefaction waves will appear. Due to this process, reconnection–associated disturbances start to propagated along the current sheet in the form of pairs of shocks. Until the reconnection electric field ceases, the shocks are detached to the reconnection site forming a teardrop–shaped region of heated and accelerated plasma, the so–called outflow regions (shaded areas in Fig. 4).

After the electric field has dropped to zero, the shocks detach from the reconnection site and continue to propagate along the current sheet (Fig. 4). These propagating shocks are causing disturbances in the ambient plasma environment, which can be measured by a satellite. In the following, we will show how these measured disturbances can be calculated by using a time–dependent Petschek–type model of transient magnetic reconnection.

3.2. MHD Treatment of the Problem for a Compressible Plasma

After the local reconnection of magnetic flux according to the Petschek-type model of magnetic reconnection, the plasma and magnetic field behavior in the outer regions can be described within the frame of ideal magnetohydrodynamics [1, 61],

$$\frac{\partial \rho}{\partial t} + \nabla \cdot (\rho \mathbf{v}) = 0\,, \tag{1}$$

$$\frac{\partial (\rho \mathbf{v})}{\partial t} + \nabla \cdot \left[\rho \mathbf{v}\mathbf{v} + p\mathbf{I} - \frac{1}{4\pi}\left(\mathbf{B}\mathbf{B} - \frac{B^2}{2}\mathbf{I} \right) \right] = 0\,, \tag{2}$$

$$\frac{\partial}{\partial t}\left(\frac{\rho v^2}{2} + \rho e + \frac{B^2}{8\pi} \right) + \nabla \cdot \left[\rho \mathbf{v} \left(\frac{v^2}{2} + e + \frac{p}{\rho} \right) + \frac{1}{4\pi} \mathbf{B} \times (\mathbf{v} \times \mathbf{B}) \right] = 0\,, \tag{3}$$

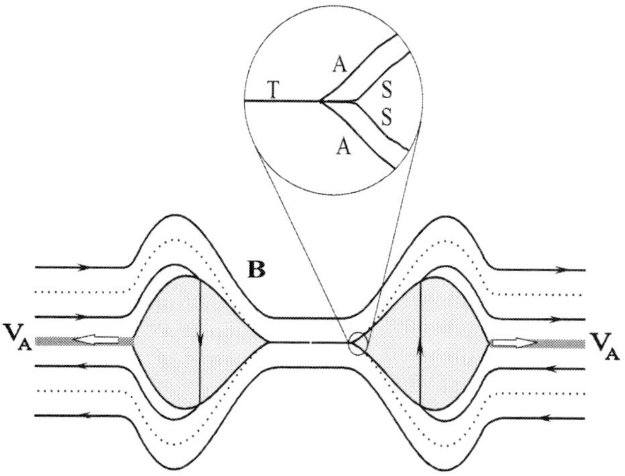

Figure 4. Shocks propagating away from the reconnection site with local Alfv én velocity. The small figure shows the internal structure of the shocks according to the general Riemann problem, where the tangential discontinuity decays into a slow shock S and an Alfvén wave A. The shaded areas are the so–called outflow regions.

$$\frac{\partial \mathbf{B}}{\partial t} + \nabla \cdot (\mathbf{B}\,\mathbf{v} - \mathbf{v}\,\mathbf{B}) = 0\,. \tag{4}$$

These equations describe the conservation of mass (Eq. 1), momentum (Eq. 2), energy (Eq. 3), and magnetic flux (Eq. 4), respectively. ρ, $\mathbf{v}$, $\mathbf{B}$, p, and e are the mass density, the plasma velocity, the magnetic field, the isotropic pressure, and the internal energy, respectively, and $\mathbf{I}$ is the unit dyadic.

These ideal MHD equations can be linearized with respect to constant background quantities, in the following denoted by superscript (0). The linearization allows for the description of all first order quantities (indicates by superscript (1)) in terms of a displacement vector ξ,

$$\mathbf{v}^{(1)} = \left(\frac{\partial}{\partial t} + \mathbf{v}^{(0)} \cdot \nabla\right) \xi\,, \tag{5}$$

$$\mathbf{B}^{(1)} = \mathbf{B}^{(0)} \cdot \nabla \xi - \mathbf{B}^{(0)} \nabla \cdot \xi\,, \tag{6}$$

$$\rho^{(1)} = -\rho^{(0)} \nabla \cdot \xi\,, \tag{7}$$

$$p^{(1)} = c_S^2 \rho^{(1)}\,, \tag{8}$$

$$P^{(1)} = -\rho^{(0)} \left[u^2 \nabla \cdot \xi + (\mathbf{v}_A \cdot \nabla)\, \mathbf{v}_A \cdot \xi\right]\,. \tag{9}$$

Here, $P^{(1)}$ is the first order total pressure, while $\mathbf{v}_A$ and c_S are the Alfvén velocity and the sound speed, respectively, and $u^2 = v_A^2 + c_S^2$.
Substituting these expressions (Eq. 5–9) into the linearized MHD equations, applying a Laplace transformation with respect to time t, and a Fourier transformation with respect to

the spatial coordinate x, the z–component of the displacement vector $\zeta(p, k, z)$ fulfills the ordinary differential equation

$$\frac{\partial^2}{\partial z^2}\zeta - q^2(p, k)\zeta = 0\,, \tag{10}$$

with

$$q^2 = \frac{p^4 + u^2k^2p^2 + v_A^2 c_s^2 k^4}{p^2u^2 + v_A^2 c_s^2 k^2}\,. \tag{11}$$

The solutions of Eq. 10 which satisfy the first order total pressure balance at $z = 0$ and vanish at infinity are

$$\zeta(p, k, z) = \frac{\tilde{L}}{L + \tilde{L}} Q(p, k)\mathrm{e}^{-qz} \qquad z > 0\,, \tag{12}$$

$$\tilde{\zeta}(p, k, z) = -\frac{L}{L + \tilde{L}} Q(p, k)\mathrm{e}^{qz} \qquad z < 0\,, \tag{13}$$

where

$$L(p, k) = -\rho^{(0)}\frac{p^2 + v_A^2 k^2}{q(p, k)}\,. \tag{14}$$

Tilde indicates the same function evaluated for the lower half space. Because of the finite boundary layer at $z = 0$, the source function $Q(p, k) = \zeta(p, k, 0) - \tilde{\zeta}(p, k, 0)$ is non–zero and can be found from the solution of the Riemann problem [1].

The expression for the displacement vector in the coordinate–time space can be found by using the so–called Cagniard–deHoop method [62, 63, 64, 59]. This method allows an analytical solution of the inverse Fourier transformation. The inverse Laplace transformation can be rewritten leading to an expression for the displacement vector as a convolution integral of the reconnection rate and an integration kernel in time, which can be solved numerically. After this procedure, the z–component of the displacement vector in coordinate–time space is found as

$$\zeta(x, z, t) = \frac{B}{\pi}\int_C \Re \frac{\tilde{L}}{L + \tilde{L}} Q(s)F(t - \tau(s))d\tau\,, \tag{15}$$

where $F(t)$ is the reconnected flux, $s = k/p$ and the integration is performed in the complex plane along a path C where the integration variable τ is real. From Eqs. 5–9 it is possible to derive the coordinate–time expressions for all relevant MHD parameters, like the pressure, the plasma density, the magnetic field and the flow velocity for a given reconnection rate. The variation of the z–component of the magnetic field, which is used in the following for the reconstruction method, is given as

$$B_z^{(1)}(x, z, t) = \frac{B^2}{\pi}\int_C \Re \frac{\tilde{L}}{L + \tilde{L}} i\, s\, Q(s)\, E(t - \tau(s))\, d\tau\,. \tag{16}$$

The expressions for the other quantities have a similar mathematical structure. In Fig. 5, the variations of the main MHD parameters are shown for fixed time ($t = 3$) in the x–z–plane. The reconnection rate is assumed to be

$$E(t) = \frac{b^2 e^2}{20} t^2 e^{-b\,t}\,,$$

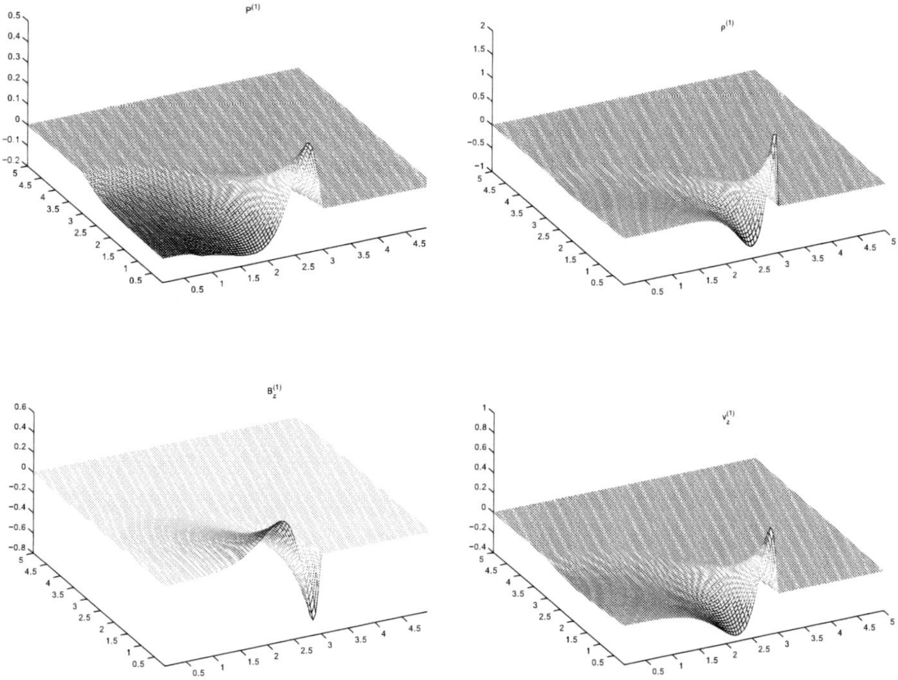

Figure 5. Variations of the total pressure (upper left panel), the plasma density (upper right panel), the z–component of the magnetic field (lower left panel), and the z–component of the flow velocity (lower right panel) for the upper half plane. The magnetic fields are $B = 1$ and $\tilde{B} = -1$, $t = 3$, and $\beta = 0.1$.

where $b = 4$.
It is possible to simulate magnetic field measurements along a satellite trajectory by using Eq. 16. Fig. 6 shows a time series of $B_z^{(1)}$ measured by a satellite fixed in space. The characteristic asymmetric bipolar variation expected for the B_z–component is clearly visible, and also the deflection of the x–component of the magnetic field. Additionally, the velocity behaves as proposed for NFTEs, with a upward flow of plasma followed by a strong flow directed downward to the plasma sheet.

4. Solution of the Inverse Problem

Since we are interested to reconstruct the reconnection rate $E(t)$ from satellite measurements of the magnetic field components, e.g., $B_z(t)$, we have to solve an inverse problem. An inverse problem can be seen as a model of a phenomenon characterized by φ, which belongs to a certain space of models P. Furthermore, let u be the observed indirect attributes of the phenomenon with $u \in U$. An operator A relates the two values by

$$A\varphi = u\,. \tag{17}$$

As a rule, the φ attributes cannot be directly observed.
The main job when solving an inverse problem is to find out, whether the chosen model is

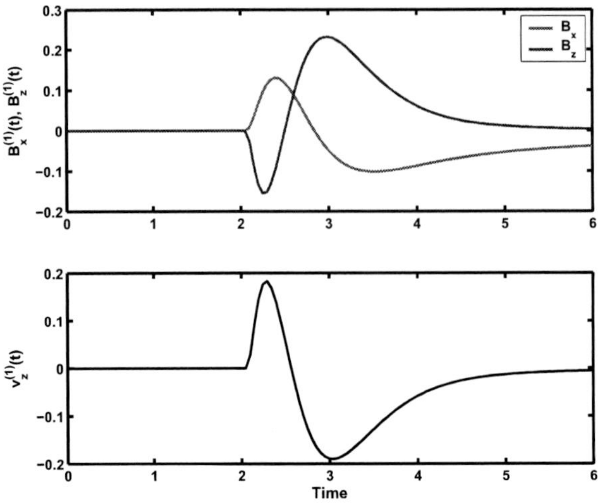

Figure 6. The normal (B_z) and the tangential (B_x) component of the magnetic field (upper panel) and the plasma flow velocity normal to the current sheet (v_z; lower panel). All characteristic features expected for NFTEs, like the bipolar variation of B_z and the anticorrelated behavior of v_z can be seen.

compatible with the experimental data. The mathematical difficulty of solving such problems is that the inverse operator A^{-1} defined throughout its domain $AP \subset U$ is not continuous. Hence there is the conventional division into the classes of well–posed and ill–posed problems. A well–posed problem, as defined by Hadamard [65], must meet the requirements that:

1. for each data set u, Eq. 17 is solvable in the entire space U,
2. the solution is unique,
3. the solution is stable, meaning that small perturbations of u result in small perturbations of the solution φ.

If a problem does not meet one or more of these criteria, the problem is considered to be ill–posed. These kind of problems can be solved by using the theory of regularization, which we apply in the following.
To reconstruct the reconnection electric field from the magnetic field, we apply a Laplace transformation to the convolution integral (16) giving

$$B_z^{(1)}(p) = K(p)E(p)\,, \tag{18}$$

where $K(p)$ represents the kernel of the convolution integral (for convenience, the factor B^2/π in front of the integral in Eq. 16 is put into $K(p)$). Thus, the reconnection electric field is given as $E(p) = B_z^{(1)}(p)/K(p)$. From this expression it is obvious that we are faced with an ill–posed inverse problem. If $K(p) \to 0$, requirement 1 is not fulfilled. Additionally, if $K(p) \sim 0$, small perturbations in the measured data $B_z^{(1)}(p)$ will cause

large perturbations in the result $E(p)$.
To avoid this problem, we introduce a regularization parameter $M(p)$ [66] so that

$$E(p) = \frac{B_z^{(1)}(p)}{K(p) + M(p)}. \tag{19}$$

The regularization operator $M(p)$ is defined as [67]

$$M(p) = \begin{cases} 0 & |p| < R_{max} \\ \infty & |p| > R_{max} \end{cases}. \tag{20}$$

This operator does not influence the electric field for small values of p, but when the functions tend to zero for large p ($p > R_{max}$), the denominator goes to infinity, so that the reconnection electric field is zero in Laplace space and large oscillations are suppressed. The value of R_{max} is found from internal parameters of the numerical Laplace transformation.

5. Reconstruction of the Reconnection Rate and Site

After the observation of FTE signatures, several attempts were made to reconstruct and analyze different features of the reconnection process involved in the generation of FTEs. Southwood [68] predicted that FTE signatures would be observed by a satellite regardless of whether or not it actually penetrates the FTE. This suggestion was verified by Farrugia et al. [69], who reproduced the magnetic field signatures outside the flux tube by considering the flow of an inviscid, incompressible plasma over a semi-circular cylinder. In a new approach, Walthour et al. [70, 71] used a method based on integral transforms for inferring the cross–sectional size, shape, and the speed of propagation of a thin, infinitely long obstacle corresponding to a flux tube.

Another approach to this topic was used by several authors [72, 73, 74] who developed a method based on the Grad–Shafranov equation to reconstruct two-dimensional space plasma structures in magnetohydrostatic equilibrium. Hu and Sonnerup [74] applied this model to two magnetopause crossings by the spacecraft AMPTE/IRM, and reconstructed magnetic field structures. Sonnerup et al. [75] used this method to give a rough estimation of the reconnection rate. However, the propagation of a reconnected flux tube can rather be described by the inertia force of the plasma and the Maxwellian tension, which is essentially a time–dependent process, and therefore this process can be hardly understood in a magnetohydrostatic approach.

Recently, different models based on multi–spacecraft measurements were developed to determine the position of the reconnection site with respect to the satellite. Wild et al. [76] used data from Cluster and Geotail positioned at the high– and the low–latitude dayside magnetopause. They consider the motion of reconnected flux tubes away from a user–defined location and compare the flux tube motion with the observed FTE–like signatures. In another approach, Fuselier et al. [77] determine the reconnection inflow velocity into the magnetosphere, and use Cluster and IMAGE data to estimate the approximate reconnection site. A third approach to reconstruct the reconnection rate, which was discussed above, was developed by Semenov et al. [67] for an incompressible plasma. It was applied to

Cluster measurements in the magnetotail by Semenov et al. [48] and Penz et al. [49]. In the following, three applications of this method are presented.

5.1. Cluster Observations on September 8th, 2002

On September 8th, 2002, a favorable constellation of multiple spacecraft and ground–based observations allowed to reconstruct the time sequence of a substorm and to model the near-Earth magnetic configuration [35]. Between 20 and 23 UT, an isolated substorm with a peak AE of about 400 nT occurred. A clear growth phase was observed after the arrival of a southward–orientated IMF after 20 UT. The auroral breakup, the intensification of a westward electrojet, and Pi2 pulsations consistently indicated the expansion phase onset at 21:18 UT in the 22–24 MLT sector. The Cluster tetrahedron was centered in the middle of the magnetotail at $[-16.7; 0.2; 4.5]$ R_e GSM. The satellites exited from the thinning plasma sheet shortly after 21:00 UT. Therefore, they were located outside of the plasma sheet at the time of interest. After 21:17 UT, a series of Earthward propagating 1 minute scale variations of the magnetic field and plasma flow components consistent with the picture of multiple NFTEs/flux ropes were observed (Fig. 7). The first NFTE appeared at about 21:17 UT and propagated Earthward at a speed v_x=625 km/s; v_y=-72 km/s (determined from timing of magnetic variations, [35]). After this NFTE, the plasma sheet continued to be thin for some 20 min until the transient plasma sheet expansions start to be observed. This is a favorable situation, because if the plasma sheet is thin, the approximation of a tangential discontinuity as an initial state is better justified.

Fig. 7 shows the measured B_x, B_z, and v_z from the four Cluster satellites during the time interval from 21:16 to 21:26 UT. The GSM magnetic field data are obtained from the fluxgate magnetometer (FGM) experiment [78] with 1 s time resolution. The O^+ moments with 4 s time resolution were measured by the Composition and Distribution Function Analyser (CODIF) of the Cluster Ion Spectrometry (CIS) experiment [79] observed at the Cluster spacecrafts. The O^+ data was only used if the O^+ density exceeded 0.005 cm^{-3}. We applied our model to the NFTEs starting at 21:21 UT, 21:22:30 UT, and 21:24 UT, ignoring the first NFTE at 21:17 UT (in which the interaction of reconnected flux tube with previously closed plasma sheet flux tubes should be more pronounced than in the following development). A comparison with the results of the theoretical model (Fig. 6) shows that the expected features for the perturbations are indeed found in the observations : the asymmetric bipolar variation of B_z, a compression of B_x, as well as a plasma flow v_z of cold O^+ ions directed to the plasma sheet are clearly visible.

5.1.1. Reconstruction of the Events on September 8th, 2002

To evaluate the integration kernel $K(p)$ we also need to know the spacecraft location with respect to the reconnection site (Fig. 8). Fortunately, the actual z–position of the neutral sheet (about +1 R_e in z–direction) is known from the modeling made in [35]. Therefore, the z–distance between the satellite and the reconnection site is approximately 3.5 R_e.

The determination of the x–distance is done by using a global minimization routine. In our time–dependent model this is possible, because the shape of the shock is changing if the x–distance from the reconnection site is increasing. We use the measured B_z component as

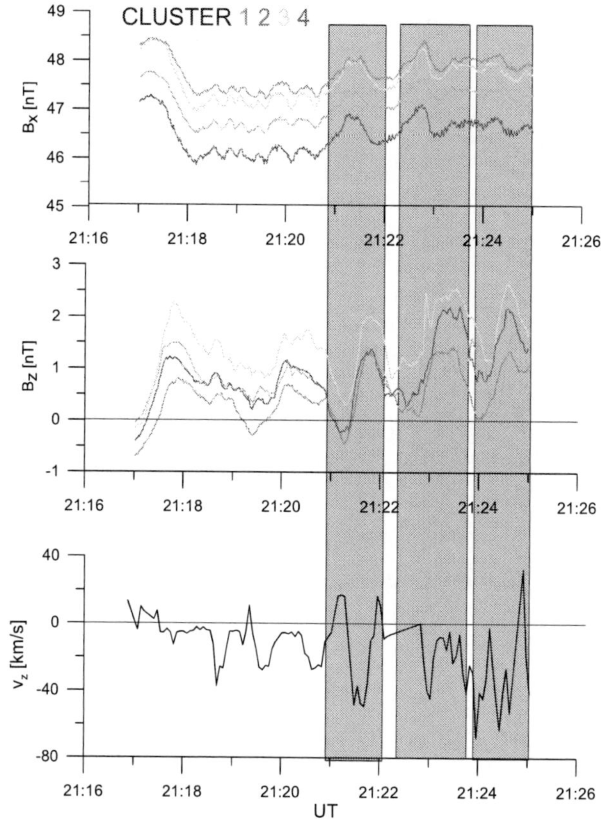

Figure 7. Cluster measurements of B_x (upper panel), B_z (middle panel), and v_z (lower panel). We analyze the NFTEs occurring at 21:21 UT, 21:22:30 UT, and 21:24 UT (shaded areas) [48].

an input data, calculate the electric field in Laplace space, which should be strictly positive, but since we do not know the x–distance it can appear negative somewhere. Therefore, we take the module of the electric field and recalculate $\tilde{B}_z$ out of it. This procedure can be summarized as

$$B_z(t) \Rightarrow B_z(p) \Rightarrow E(p) \Rightarrow E(t) \Rightarrow |E(t)| \Rightarrow \tilde{B}_z(t)\,.$$

Then we minimize the difference between B_z(t) and $\tilde{B}_z$(t) with a least square approach in order to find the x–distance as the value of x where the difference between the initial and the reconstructed magnetic field has a minimum. We limit the search to distances less than 35 R_e, which corresponds to the region of the near Earth neutral line (NENL), where reconnection most likely takes place. The local velocity of the disturbances is determined by using multipoint timing analysis [80], giving about 700 km/s. We assume that this velocity is approximately the Alfvén velocity. Additionally, this analysis shows that the propagation velocity is directed mainly Earthward in x–direction with a small y–component, which is a preferable configuration for our 2-D model. Now we know all quantities used for the

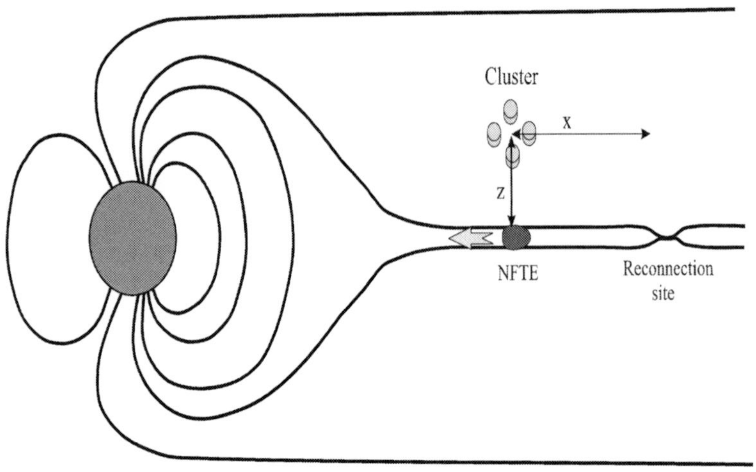

Figure 8. The position of the Cluster satellites with respect to the reconnection site. The z–distance is found from the modeling done by Sergeev et al. [35], while the x–distance is reconstructed by using a minimization routine.

normalization. Thus, we are able to analyze the events by using the method presented in the previous sections.

The reconnection electric field reconstructed for the NFTE starting at 21:21 UT in Fig. 7 is 1.1 mV/m over a time period of about 50 s from C1. C2 and C3 give a reconnection rate of 1.0 mV/m, while C4 gives 1.1 mV/m (Fig. 9).

The distance x between C1 and the reconnection site is 8.6 R_e, so that the reconnection site is located at 24.7 R_e in the magnetotail (Fig. 10). For C2, $x = 8.6$, giving 24.9 R_e as the location of the reconnection site. From C3 x is found to be 8.4, giving the reconnection site at 24.6 R_e. C4 gives $x = 7.6$, corresponding to a reconnection site at 24.3 R_e. One can see that the reconstructed amplitude of the reconnection electric field and the reconnection site are consistent among the four satellites. For example, C3 is located about 0.5 Re lower than C2, therefore the observed amplitude in the magnetic field perturbation at C3 is 1.7 nT compared with 1.3 nT at C2. Yet, the reconstructed amplitude of the electric field is quite similar, i.e., 1.0 mV/m from C3 and from C2. The same can be found for the x–distance, where C4 is located closest to the reconnection site, and therefore the model gives the smallest distance between the satellite and the reconnection site. Therefore our model shows that the observed features at all four satellites are associated with the same reconnection event.

Analysis of the event starting at 21:22:30 UT was done only with C2 and C4, since the other two satellites did not observe pronounced NFTE–like signatures. The reconnection rates are found to be 0.95 and 1.0 mV/m for C2 and C4, respectively. C2 gives $x = 8.6$, leading to a distance of 24.9 R_e for the reconnection site. From C4, x is found to be 8.25 R_e, giving a distance of 24.8 R_e in the magnetotail. For the event starting at 21:24 UT, data from C2, C3, and C4 are used. The reconnection rate is found to be 0.95 mV/m (C2), 0.91 mV/m (C3), and 0.9 mV/m (C4). According to these data, the reconnection site is located

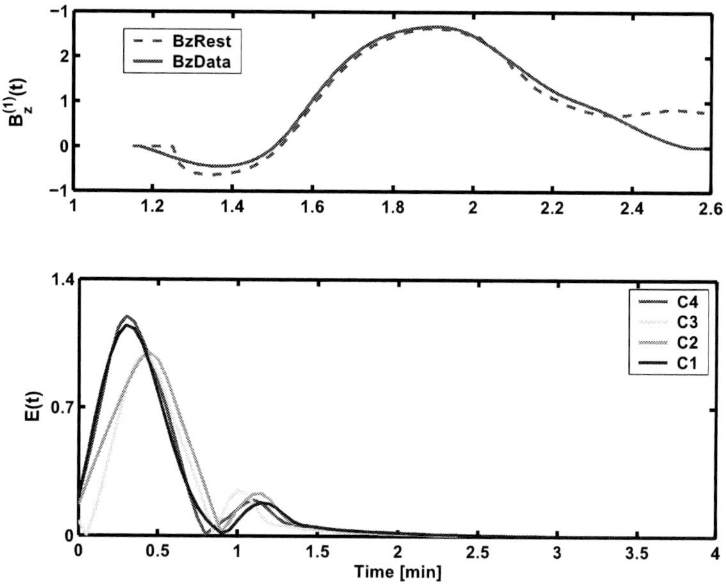

Figure 9. The lower panel shows the reconstructed reconnection rate from the four Cluster satellites, which has an amplitude of 1–1.1 mV/m and a duration of about 50 s. The upper panel shows a comparison between the measured $B_z^{(1)}$ and the restored $B_z^{(1)}$ from the reconstructed reconnection rate for C4.

at 24.9 R$_e$ (C2), 23.9 R$_e$ (C3), and 24.5 R$_e$ (C4) in the magnetotail.

6. Conclusion

We present a model to reconstruct the reconnection rate and the reconnection site from satellite measurements in a compressible plasma. As already predicted [48], the reconnection rate is smaller than inferred from the incompressible model. Also the distance between the reconnection site and the satellite decreased, while the duration of the pulses increase. All these features are consistent with the qualitative estimates done by Semenov et al. [48]. However, a significant reduction of the noise resulting from the solution of the inverse problem is achieved compared with the incompressible model [48]. Also the variations of the reconstructed reconnection rate and site between the satellites decreased.

Several simplifying assumptions were made to allow a treatment of the problem within an analytical model:

- x-distance: The x–distance between the satellite and the reconnection site is determined by using a minimization routine. There exists the possibility that more that one minimum occurs, meaning that there is no single solution for the problem considered. To avoid this problem, we applied our method only to the range of x–distances, where reconnection most likely takes place, namely to distances less than 35 R$_e$, and run it with different starting points.

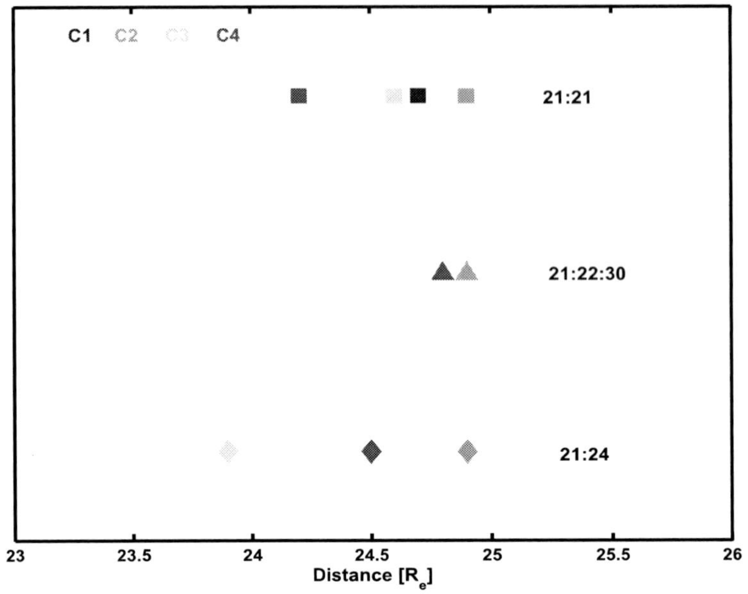

Figure 10. Location of the reconnection site reconstructed for the NFTE starting at 21:21 UT (squares), 21:22:30 UT (triangles), and 21:24 UT (diamonds) from the Cluster satellites.

- Homogeneous background density : We assume that there is a homogeneous background density in the magnetotail. If the density changes significantly between the point of observation and the starting point of the disturbances, this will influence the propagation speed of the perturbations and may modify the results.

- Alfvén velocity: We assume that the perturbations in the magnetic field are moving approximately with Alfvén velocity. If the Alfvén velocity is not constant between the satellite and the reconnection site, the results will change. Since the Alfv´en velocity is used for the normalization of the length scales, a variation of the Alfv´en velocity will also give a variation of the spatial distances. If the Alfv´en velocity decreases, also the length scales will decrease.

However, we found that the variations observed by the Cluster spacecrafts outside of the plasma sheet boundary are consistent with the impulsive model of magnetic reconnection used in this work. The characteristic bipolar variation of the B_z–component and the anticorrelated behavior of the perpendicular plasma flow velocity are reproduced. Application to NFTEs occurring on September 8th, 2002, gives an amplitude of the reconnection electric field in the range of 0.9–1.2 mV/m. This amplitude of the reconnection electric field is consistent with estimations of the magnetotail reconnection rate obtained from ground–based measurements [81, 82]. The time duration of the reconnection pulse is in the order of 50 s, while the reconnection site is located at about 23.9–24.9 R_e tailwards.

Acknowledgments

The authors thank A. Runov for fruitful discussions. This work is supported by the RFBR grants No. 04–05–64935 and No. 03–05–20012 BNTS, by the Austrian "Fonds zur Förderung der wissenschaftlichen Forschung" under project P17099–N08, by project No. I.12/04 from the "Österreichischer Austauschdienst", and by INTAS Grant Nr. 03–51–3738. V.S.S. acknowledges financial support from the Technical University Graz and the hospitality of the Institut für Theoretische Physik during scientific visits to Graz. T.P. acknowledges support by the "Österreichische Forschungsgemeinschaft" and by a Paul-Urban scholarship from the Institute of Physics of the University of Graz during visits in St. Petersburg. V.V.I. acknowledges support by INTAS "Fellowship Grant for Young Scientists" Nr. 04–83–3816, and by the Nansen community. Also acknowledged is support by the Austrian Academy of Sciences, "Verwaltungsstelle für Auslandsbeziehungen".

References

[1] Semenov, V. S.; Heyn, M. F.; Ivanov, I. B. *Phys. Plasmas* 2004, 11, 62–70.

[2] Sweet, P. A. In *Electromagnetic Phenomena in Cosmic Physics*; Lehnert, B.; Ed.; Cambridge University Press: London, UK, 1958, pp 123–134.

[3] Parker, E. N. *Astrophys. J. Suppl.* 1963, 8, 177–211.

[4] Petschek, H. E. In *The Physics of Solar Flares*; Hess, W. N.; Ed.; NASA Spec. Publ.: Washington, DC, USA, 1964; SP–50, pp 425–440.

[5] Erkaev, N. V.; Semenov, V. S.; Jamitzky, F. *Phys. Rev. Lett.* 2000, 84, 1455–1458.

[6] Erkaev, N. V.; Semenov, V. S.; Alexeev, I. V.; Biernat, H. K. *Phys. Plasmas* 2001, 8, 4800–4809.

[7] Birn, J.; Drake, J. F.; Shay, M. A.; Rogers, B. N.; Denton, R. E.; Hesse, M.; Kuznetsova, M.; Ma, Z. W.; Bhattacharjee, A.; Otto, A.; Pritchett, P. L. *J. Geophys. Res.* 2001, 106, 3715–3720.

[8] Phan, T. D. et al. *Nature* 2000, 404, 848–850.

[9] Priest, E.; Forbes, T. *Magnetic Reconnection: MHD Theory and Applications*; Cambridge University Press: Cambridge, UK, 2000.

[10] Paré, V. K. In *Magnetic reconnection in space and laboratory plasmas*; Hones, E. W.; Ed., Geohysical Monograph: Washington, DC, USA, 1984; 30, pp 341–346.

[11] Yamada, M.; Levinton, F. M.; Pomphrey, N.; Budny, R.; Manickam, J.; Nagayama, Y. *Phys. Plasmas* 1994, 1, 3269–3276.

[12] Yamada, M.; Ji, H.; Hsu, S.; Carter, T.; Kulsrud, R.; Bretz, N.; Jobes, F.; Ono, Y.; Perkins, F. *Phys. Plasmas* 1997, 4, 1936–1944.

[13] Parker, E. N. *J. Geophys. Res.* 1957, 62, 509–520.

[14] Scholer, M. *Lecture Notes in Physics* 2003, 612, 9–27.

[15] Biernat, H. K.; Semenov, V. S.; Penz, T.; Miklenic, C.; Veronig, A.; Hanslmeier, A.; Vrsnak, B.; Heyn, M. F.; Ivanov, I. B.; Ivanova, V. V.; Kiehas, S.; Langmayr, D. *Hvar Obs. Bullet.* 2006, in press.

[16] Uzdensky, D. A. *Astrophys. J.* 2003, 587, 450–457.

[17] Axford, W. I.; McKenzie, J. F. *Astrophys. Space Sci.* 1996, 243, 1–4.

[18] Mullan, D. J. In *The M–type stars*; Johnson, J. R.; Querci, F. R.; Ed.; NASA Spec. Publ: Washington, DC, USA 1986; SP–492, pp 455–479.

[19] Verbunt, F. *Space Sci. Rev.* 1982, 32, 379–404.

[20] Cowley, S. W. H. In *Magnetic Reconnection in Space and Laboratory Plasmas*; Hones, E. W.; Ed.; Geohysical Monograph: Washington, DC, USA, 1984; 30, pp 375–378.

[21] Russell, C. T.; Elphic, R. C. *Space Sci. Rev.* 1978, 22, 681–715.

[22] Haerendel, G.; Paschmann, G.; Sckopke, N.; Rosenbauer, H.; Hedgecock, P. C. *J. Geophys. Res.* 1978, 83, 3195–3216.

[23] Rijnbeek, R. P.; Cowley, S. W. H.; Southwood, D. J.; Russell, C. T. *J. Geophys. Res.* 1984, 89, 786–800.

[24] Kawano, H.; Russell, C. T. *J. Geophys. Res.* 1997, 102, 11307–11314.

[25] Twitty, C.; Phan, T. D.; Paschmann, G.; Lavraud, B.; Rème, H.; Dunlop, M. *Geophys. Res. Lett.* 2004, 31, L19808, doi:10.1029/2004GL020646.

[26] Frey, H. U.; Phan, T. D.; Fuselier, S. A.; Mende, S. B. *Nature* 2003, 426, 533–537.

[27] Walker, R. J.; Ogino, T.; Raeder, J.; Ashour–Abdalla, M. J. *Geophys. Res.* 1993, 98, 17235–17250.

[28] Lopez, R. E.; Benitez–Marquez, E.; Wiltberger, M. J.; Lyon, J. G.; Figueroa, R. *Adv. Space Res.* 2003, 31, 1167–1176.

[29] Sergeev, V. A.; Semenov, V. S.; Sidneva, M. V. *Planet. Space Sci.* 1987, 35, 1199–1212.

[30] Sergeev, V. A.; Elphic, R. C.; Mozer, F. S.; Saint–Marc, A.; Sauvaud, J. A. *Planet. Space Sci.* 1992, 40, 1551–1572.

[31] Ieda, A.; Machida, S.; Mukai, T.; Saito, Y.; Yamamoto, T.; Nishida, A.; Terasawa, T.; Kokubun, S. J. *Geophys. Res.* 1998, 103, 4453–4465.

[32] Slavin, J. A.; Lepping, R. P.; Gjerloev, J.; Fairfield, D. H.; Hesse, M.; Owen, C. J.; Moldwin, M. B.; Nagai, T.; Ieda, A.; Mukai, T. J. *Geophys. Res.* 2003, 108, 1015, doi:10.1029/2002JA009557.

[33] Nagai, T.; Shinohara, I.; Fujimoto, M.; Hoshino, M.; Saito, Y.; Machida, S.; Mukai, T. J. *Geophys. Res.* 2001, 106, 25929–25950.

[34] Runov, A.; Nakamura, R.; Baumjohann, W.; Treumann, R. A.; Zhang, T. L.; Volwerk, M.; Vörös, Z.; Balogh, A.; Glassmeier, K.–H.; Klecker, B.; Rème, H.; Kistler, L. *Geophys. Res. Lett.* 2003, 30, 1579, doi:10.1029/2002GL016730.

[35] Sergeev, V. A.; Kubyshkina, M. V.; Baumjohann, W.; Nakamura, R.; Amm, O.; Pulkkinen, T.; Angelopoulos, V.; Mende, S. B.; Klecker, B.; Nagai, T.; Sauvaud, J.–A.; Slavin, J. A.; Thomsen, M. F. *Ann. Geophys.* 2005, in press.

[36] Kistler, L.; Mouikis, C.; Moebius, E.; Klecker, B.; Sauvaud, J.–A.; Rème, H.; Korth, A.; Marocci, M.; Lundin, R.; Parks, G.; Balogh, A. J. *Geophys. Res.* 2005, 110, A06213, doi:10.1029/2004JA010653.

[37] Wygant, J. R.; Cattell, C. A.; Lysak, R.; Song, Y.; Dombeck, J.; McFadden, J.; Mozer, F. S.; Carlson, C.; Parks, G.; Lucek, E. A.; Balogh, A.; Andre, M.; Rème, H.; Hesse, M.; Mouikis, C. J. *Geophys. Res.* 2005, submitted.

[38] Cattell, C.; Dombeck, J.; Wygant, J.; Drake, J. F.; Swisdak, M.; Goldstein, M. L.; Keith, W.; Fazakerley, A.; Andre, M.; Lucek, E.; Balogh, A. J. *Geophys. Res.* 2005, 110, A01211, 10.1029/2004JA010519.

[39] Nakamura, R.; Baumjohann, W.; Nagai, T.; Fujimoto, M.; Mukai, T.; Klecker, B.; Treumann, R. A.; Balogh, A.; Rème, H.; Sauvaud, J.–A.; Kistler, L.; Mouikis, C.; Owen, C. J.; Fazakerley, A.; Dewhurst, J.; Bogdanova, Y. J. *Geophys. Res* 2004, 109, A05204, doi:10.1029/2003JA010174.

[40] Alexeev, I. V.; Owen, C. J.; Fazakerley, A. N.; Runov, A.; Dewhurst, J.; Balogh, A.; Rème, H.; Klecker, B.; Kistler, L. *Geophys. Res. Lett.* 2005, 32, L03101, doi:10.1029/2004GL021420.

[41] Asano, Y.; Nakamura, R.; Baumjohann, W.; Runov, A.; Vörös, Z.; Volwerk, M.; Zhang, T. L. *Geophys. Res. Lett.* 2005, 32, L03108, doi:10.1029/2004GL021834.

[42] Treumann, R. A.; Jaroschek, C. H.; Nakamura, R.; Runov, A.; Scholer, M. *COSPAR Colloquia Series*, 2005, in press.

[43] Eriksson, S.; Oieroset, M.; Baker, D.; Mouikis, C.; Vaivads, A.; Dunlop, M.; Rème, H.; Ergun, R.; Balogh, A. J. *Geophys. Res.* 2004, 109, A05212, doi:10.1029/2003JA010534.

[44] Slavin, J. A.; Lepping, R.; Gjerloev, J.; Fairfield, D.; Acuna, M.; Goldstein, M.; Balogh, A.; Dunlop, M.; Kivelson, M.; Khurana, K.; Fazakerley, A.; Owen, C. J.; Rème, H.; Bosqued J. *Geophys. Res. Lett.* 2003, 30, 1362, doi:10.1029/2003GL016411.

[45] Zong, Q.–G.; Fritz, T. A.; Pu, Z. Y.; Fu, S. Y.; Baker, D. N.; Zhang, H.; Lui, A. T.; Vogiatzis, I.; Glassmeier, K.–H.; Korth, A.; Daly, P. W.; Balogh, A.; Rème, H. *Geophys. Res. Lett.* 2004, 31, L18803, doi:10.1029/2004GL020692.

[46] Slavin, J. A.; Owen, C. J.; Dunlop, M. W.; Boralev, E.; Moldwin, M. B.; Sibeck, D. G.; Tanskanen, E.; Goldstein, M. L.; Fazaderley, A.; Balogh, A.; Lucek, E.; Richter, I.; Rème, H.; Bosqued, J. *Geophys. Res. Lett.* 2003, 30, 2208, doi:10.1029/2003GL018438.

[47] Owen, C. J.; Slavin, J. A.; Fazakerley, A. N.; Dunlop, M. W.; Balogh, A. *Geophys. Res. Lett.* 2005, 32, L03104, 10.1029/2004GL021767.

[48] Semenov, V. S.; Penz, T.; Ivanova, V. V.; Sergeev, V. A.; Biernat, H. K.; Nakamura, R.; Heyn, M. F.; Kubyshkin, I. V.; Ivanov, I. B. *J. Geophys. Res.* 2005, 110, A11217, doi:10.1029/2005JA011181.

[49] Penz, T.; Semenov, V. S.; Ivanova, V. V.; Biernat, H. K.; Sergeev, V. A.; Nakamura, R.; Kubyshkin, I. V.; Ivanov, I. B.; Heyn, M. F. *Adv. Space Res.* 2006, in press.

[50] Nakamura, R.; Baumjohann, W.; Mouikis, C.; Kistler, L. M.; Runov, A.; Volwerk, M.; Asano, Y.; Vörös, Z.; Zhang, T. L.; Klecker, B.; Rème, H.; Balogh, A. *Geophys. Res. Lett.* 2004, 31, L09804, doi:10.1029/2004GL019558.

[51] Nakamura, R.; Baumjohann, W.; Klecker, B.; Bogdanova, Y.; Balogh, A.; Rème, H.; Bosqued, J.-M.; Dandouras, I.; Sauvaud, J.–A.; Glassmeier, K.–H.; Kistler, L.; Mouikis, C.; Zhang, T. L.; Eichelberger, H.; Runov, A. *Geophys. Res. Lett.* 2002, 29, 1942, doi:10.1029/2002GL015763.

[52] Volwerk, M.; Nakamura, R.; Baumjohann, W.; Treumann, R. A.; Runov, A.; Vörös, Z.; Zhang, T. L.; Asano, Y.; Klecker, B.; Richter, I.; Balogh, A.; Rème, H. *J. Geophys. Res.* 2003, 108, 1429, doi:10.1029/2003JA010155.

[53] Vörös, Z.; Baumjohann, W.; Nakamura, R.; Runov, A.; Zhang, T. L.; Volwerk, M.; Eichelberger, H.; Balogh, A.; Horbury, T. S.; Glassmeier, K.–H.; Klecker, B.; Rème, H. *Ann. Geophys.* 2003, 21, 1955–1964.

[54] Grocott, A.; Yeoman, T.; Nakamura, R.; Cowley, S. W. H.; Frey, H.; Rème, H.; Klecker, B. *Ann. Geophys.* 2004, 22, 1061–1075.

[55] Nakamura, R.; Amm, O.; Laakso, H.; Draper, N.; Lester, M.; Grocott, A.; Klecker, B.; McCrea, I.; Balogh, A.; Rème, H.; Andre, M. *Ann. Geophys.* 2005, 99, 553–566.

[56] Biernat, H. K.; Heyn, M. F.; Semenov, V. S. *J. Geophys. Res.* 1987, 92, 3392–3396.

[57] Semenov, V. S.; Kubyshkin, I. V.; Lebedeva, V. V.; Rijnbeek, R. P.; Heyn, M. F.; Biernat, H. K.; Farrugia, C. J. *Planet. Space Sci.* 1992, 40, 63–87.

[58] Semenov, V. S.; Kubyshkin, I. V.; Rijnbeek, R. P.; Biernat, H. K. In *Physics of magnetic reconnection in high–temperature plasmas* ; Ugai, M.; Ed.; Research Signpost: Trivandrum, India, 2004, pp 35–68.

[59] Heyn, M. F.; Semenov, V. S. *Phys. Plasmas* 1996, 3, 2725–2741.

[60] Heyn, M. F.; Biernat, H. K.; Rijnbeek, R. P.; Semenov, V. S. *J. Plasma Physics* 1988, 40, 235–252.

[61] Akhiezer, A. I.; Akhiezer, I. A.; Polovin, R. V.; Sitenko, A. G.; Stepanov, K. N. *Plasma Electrodynamics; Pergamon Press* : Oxford, UK, 1975.

[62] Lamb, H. *Philos. Trans. R. Soc. London Ser. A* 1904, 203, 1–42.

[63] Cagniard, L. *Reflexion et Refraction des Ondes Seismiques Progressive* ; Gauthier–Villard: Paris, France, 1939; English translation by Flinn, E. A.; Dix, C. H. *Reflection and refraction of progressive seismic waves* ; McGraw–Hill: New York, USA, 1962.

[64] deHoop, A. T. *Appl. Sci. Res. B* 1960, 8, 349–356.

[65] Hadamard, J. *Le problème de Cauchy et les équations aux dérivées partielles Lineaires Hyperboliques*; Hermann: Paris, France, 1932.

[66] Tikhonov, A. N.; Arsenin, V. Y. *Solutions of ill-posed problems* ; John Wiley: New York, USA, 1977.

[67] Semenov, V. S.; Penz, T.; Heyn, M. F.; Ivanov, I. B.; Kubyshkin, I. V.; Biernat, H. K.; Ivanova, V. V. In *Solar–planetary relations* ; Biernat, H. K.; Lammer, H.; Vogl, D. F.; Mühlbachler, S.; Eds.; *Research Signpost*: Trivandrum, India, 2005, pp 261–302.

[68] Southwood, D. J. *Adv. Space Res.* 1985, 5, 7–14.

[69] Farrugia, C. J.; Elphic, R. C.; Southwood, D. J.; Cowley, S. W. H. *Planet. Space Sci.* 1987, 35, 227–240.

[70] Walthour, D. W.; Sonnerup, B. U. Ö.; Paschmann, G.; Lühr, H.; Klumpar, D.; Potemra, T. J. *Geophys. Res.* 1993, 98, 1489–1504.

[71] Walthour, D. W.; Sonnerup, B. U. Ö.; Elphic, R. C.; Russell, C. T. *J. Geophys. Res.* 1994, 99, 8555–8563.

[72] Hau, L.–N.; Sonnerup, B. U. Ö. *J. Geophys. Res.* 1999, 104, 6899–6918.

[73] Hu, Q.; Sonnerup, B. U. Ö. *Geophys. Res. Lett.* 2001, 28, 467–470.

[74] Hu, Q.; Sonnerup, B. U. Ö. *J. Geophys. Res.* 2003, 108, 1011, doi:10.1029/2002JA009323.

[75] Sonnerup, B. U. Ö.; Hasegawa, H.; Paschmann, G. *Geophys. Res. Lett.* 2004, 31, L11803, doi:10.1029/2004GL020134.

[76] Wild, J. A.; Milan, S. E.; Cowley, S. W. H.; Bosqued, J. M.; Rème, H.; Nagai, T.; Kokubun, S.; Saito, Y.; Mukai, T.; Davies, J. A.; Cooling, B. M. A.; Balogh, A.; Daly, P. W. *Ann. Geophys.* 2005, 23, 445–460.

[77] Fuselier, S. A.; Trattner, K. J.; Petrinec, S. M.; Owen, C. J.; Rème, H. *J. Geophys. Res.* 2005, 110, A06212, doi:10.1029/2004JA010805.

[78] Balogh, A. et al. *Ann. Geophys.* 2001, 19, 1207–1217.

[79] Rème, H. et al. *Ann. Geophys.* 2001, 19, 1303–1354.

[80] Harvey, C. C. In *Analysis methods for multi–spacecraft data* ; Paschmann, G.; Daly, P. W.; Eds.; ISSI SR–001, 1998.

[81] Blanchard, G. T.; Lyons, L. R.; de la Beaujardière, O.; Doe, R. A.; Mendillo, M. J. *Geophys. Res.* 1996, 101, 15265–15276.

[82] Østgaard, N.; Moen, J.; Mende, S. B.; Frey, H. U.; Immel, T. J.; Gallop, P.; Oksavik, K.; Fujimoto, M. *Ann. Geophys.* 2005, 23, 123–134.

In: Perspectives in Mathematical Physics
Editor: Gregory D. Monteggo
ISBN: 978-1-61122-984-4

MODELLING OF SHOCK WAVES AND CLUSTERING IN HYDRODYNAMIC SIMULATIONS OF GRANULAR GASES

José A. Carrillo[1]* *and Clara Salueña*[2]†
[1]ICREA (Institució Catalana de Recerca i Estudis Avançats)
and Departament de Matemàtiques, Universitat
Autònoma de Barcelona, E-08193 Bellaterra, Spain
[2]Departament de Enginyeria Mecànica-ETSEQ,
Universitat Rovira i Virgili, E-43007 Tarragona, Spain

Abstract

We show that 2D clustering with constant restitution coefficient is qualitatively well described by a compressible Navier-Stokes like system. Shock capturing highly accurate methods are needed to resolve well shock waves in the cluster formation phase. Numerical results agree with qualitative predictions on clustering.

1 Introduction

In this work, we try to clarify the potential use of hydrodynamic descriptions in the modelling of different phenomena in rapid granular flows. The physical approach to granular flows lies in describing the collective behavior of a huge set of small grains (typically approximated by homogeneous small spheres) in which particles lose a part of their kinetic energy in each collision. Rapid granular flows are characterized by the fact that the mean free path is much larger than the typical particle size.

This kind of granular flows can be modelled following classical statistical mechanics [CIP94] by Boltzmann-type equations [LSJC84, JR85, BCG00, GS95], in which one follows the evolution of the number density of particles in phase space. Inelastic collisions between particles govern the dissipation of kinetic energy by the assumption that only a fraction of the normal component of the relative impact velocity is kept after the collision. This factor is called the restitution coefficient and may depend on the normal impact velocity or local temperature [BP100, BP200, GSBMSS98, BCG00, RPBS99].

*E-mail address: carrillo@mat.uab.es
†E-mail address: clara.saluena@urv.net

In the absence of external forces and/or boundary sources, particles will lose their kinetic energy by successive collisions and, in a completely homogeneous situation, all particles will eventually move at the same velocity. Homogeneous self-similar solutions explicitly supply the typical rate of energy dissipation and the general trend towards the unique equilibrium in homogeneous situations: the Dirac delta distribution in velocity. These similarity solutions, called homogeneous cooling states, have played a significant role in recent advances of hydrodynamic descriptions [BDKS98].

Force-free granular gases often evolve following complicated dynamics due to small initial inhomogeneities in the system. In fact, the appearance of clusters and vortices is well reported in the physical literature on the subject, see [BMSS02, HM03, GZ93] and references therein. Until then, the energy dissipation as a function of time closely follows the classical Haff's law [Haf83, BE98], which results in a decay of order t^{-2} for constant restitution coefficient.

Hydrodynamic descriptions have been derived from Boltzmann equations for granular gases by following standard moment methods or expansions around either local Maxwellians or homogeneous cooling states. There is a huge set of hydrodynamic descriptions for granular gases in the literature, depending on the assumptions on the restitution coefficient, the type of expansion and the equation of state considered [JR85, GS95, SG98, BDKS98].

In principle, kinetic or hydrodynamic descriptions will cease to be valid approaching close-packing situations, in which the mean free path is not meaningful at all, or due to inelastic collapse phenomena [DLK95, GSBMSS98, LM98]. Despite of this fact, such descriptions have recently been used in situations far from their supposed limits of validity, achieving good agreement with experiments [BSSS99]. For instance, shock waves in granular gases [RBSS02, BMSS02, SR04, KGSD00] or clustering [BSSP04, HM03] have been reported to be well described by compressible Navier-Stokes like systems.

However, numerical methods used to solve the hydrodynamic equations in these cases [RBSS02, BMSS02, SR04, BSSP04, HM03] were not well adapted to the difficulties one finds in real applications. Typically, granular flows become supersonic in laboratory conditions; nearly zero density regions coexist with almost close-packing ones and steep gradient regions appear in density, velocity and temperature. Therefore, shock-capturing methods should be used to describe the overall behavior of the system while a high-order accuracy would be desirable for a smooth definition of the profiles. Moreover, the dependence of viscosity and thermal conductivity on the volume fraction, typical of granular gases, leads to divergences close to vacuum as well as in the close-packing limit.

In this work, we use existing WENO (weighted essentially non oscillatory) reconstruction methods [JS96, Shu98, SO88] to accurately compute the fluxes for the Euler part of the Navier-Stokes equations for granular gases, while using simple centered second order finite differences to approximate high order terms in the Navier-Stokes equations. Euler solvers have already been proposed in the literature using either ENO [KGSD00] or Power-ENO [SM04] methods to analyze shock waves in granular gases without taking into account the Navier-Stokes terms. We will show that clustering is well described by the Navier-Stokes equations, based on their numerical resolution.

The next section is devoted to explain in detail the hydrodynamic descriptions for granular gases with constant restitution coefficient, supplied with an equation of state capable to account for dense material. The final section is devoted to analyze the numerical method and to show the numerical results for simple 1D sedimentation profiles and 2D clustering. The expressions of the eigenvectors and eigenvalues needed to obtain decompositions of the local fluxes for general equations of state are included in the appendix.

2 Euler and Navier-Stokes Equations for Granular Gases

Let us analyze the dynamics of a dilute gas of small homogeneous perfect spheres of diameter $\sigma > 0$ undergoing inelastic collisions, i.e., a granular gas. Following the point of view of the kinetic theory, given (x, v_1), $(x - \sigma n, v_2)$ the states of two particles before the collision, where $n \in S^2$ is the unit vector along the center of both spheres, the postcollisional velocities are found by assuming that the total momentum is preserved while a fraction of the normal relative velocity is lost, that is,

$$n \cdot (v_1' - v_2') = -e((v_1 - v_2) \cdot n)$$

where $0 < e \leq 1$ is called the restitution coefficient. Using both ingredients, we construct the postcollisional velocities as

$$\begin{aligned} v_1' &= \frac{1}{2}(v_1 + v_2) + \frac{V'}{2} \\ v_2' &= \frac{1}{2}(v_1 + v_2) - \frac{V'}{2} \end{aligned}$$

where $V' = V - (1+e)(V \cdot n)n$, $V = v_1 - v_2$ and $V' = v_1' - v_2'$. Let us denote by v_1^* and v_2^* the precollisional velocities corresponding to v_1 and v_2. Let us remark that the coefficient of restitution can be a function of the normal relative velocity $|(v_1 - v_2) \cdot n|$. Following the standard procedures of kinetic theory [CIP94], we derive the Boltzmann-Enskog equation for inelastic hard spheres,

$$\frac{\partial f}{\partial t} + (v_1 \cdot \nabla_x) f = \sigma^2 Q(f, f) \tag{1}$$

where the collision operator is given by

$$\begin{aligned} Q(f,f) &= G(\nu) \int_{\mathbb{R}^3} \int_{S^2} ((v_1 - v_2) \cdot n)_+ \left[\frac{1}{e} J f(t, x, v_1^*) f(t, x + \sigma n, v_2^*) \right. \\ & \qquad \left. - f(t, x, v_1) f(t, x - \sigma n, v_2) \right] dn\, dv_2. \end{aligned}$$

The moments of the probability distribution $f(t, x, v_1)$ in phase space give the macroscopic quantities: the number density

$$\rho(t, x) = \int_{\mathbb{R}^3} f(t, x, v_1)\, dv_1,$$

the velocity field

$$U(t, x) = \frac{1}{\rho} \int_{\mathbb{R}^3} v_1 f(t, x, v_1)\, dv_1,$$

and the granular temperature

$$\theta(t,x) = \frac{1}{3\rho}\int_{\mathbb{R}^3} |v_1 - U(t,x)|^2 f(t,x,v_1)\,dv_1.$$

This Boltzmann equation includes dense gas effects in the form of the correlation function $G(\nu)$, where ν is the 3D volume fraction, i.e., $\nu = \frac{\pi}{6}\rho\sigma^3$. $G(\nu)$ is typically chosen in such a way that diverges when ν approaches the 3D maximal packing fraction, which we take as $\nu_{max} = 0.65$. Different formulas are proposed for $G(\nu)$ in the literature; we implement the one used in the experiments and simulations in [BSSS99, BMSS02], that is,

$$G(\nu) = \frac{\nu}{1 - \left(\frac{\nu}{\nu_{max}}\right)^{\frac{4}{3}\nu_{max}}}. \tag{2}$$

By different expansion methods [JR85, GS95, BP100, BP200, BCG00], the hydrodynamic equations are obtained for the granular gas,

$$\begin{aligned} \frac{\partial\rho}{\partial t} + \mathrm{div}(\rho U) &= 0 \\ \rho\left(\frac{\partial U}{\partial t} + (U\cdot\nabla)U\right) &= \mathrm{div}(P) + \rho F \\ \frac{3}{2}\rho\left(\frac{\partial\theta}{\partial t} + (U\cdot\nabla)\theta\right) &= -\mathrm{div}(q) + P : E - \gamma \end{aligned} \tag{3}$$

representing the evolution of number density of particles $\rho(t,x)$, velocity field $U(t,x) = (u,v,w)(t,x) = (U_1,U_2,U_3)(t,x)$ and granular temperature $\theta(t,x)$. P is the pressure tensor, given by

$$P_{ij} = \left[-p + \left(\lambda - \frac{2}{3}\mu\right)\sum_i E_{ii}\right]\delta_{ij} + 2\mu E_{ij}$$

with $E_{ij} = \frac{1}{2}\left(\frac{\partial U_i}{\partial x_j} + \frac{\partial U_j}{\partial x_i}\right)$. The principal pressure is related to the density and the temperature through the equation of state $p = p(\rho,\theta)$, which depends on the particular assumptions on the interactions between particles at the kinetic description. The vector $q = -\chi\nabla\theta$ models the heat flux. The viscosities λ and μ, the thermal conductivity χ and the cooling coefficient γ are density and temperature dependent. The dependence on density is expressed through the volume fraction ν occupied by the grains; the dependence on temperature is of a power type. We refer to [GS95, JR85, SG98] for explicit expressions of these coefficients for constant restitution, and to [BP100, BP200] for viscoelastic particles and different interaction mechanisms. The vector field $F(t,x)$ represents an external force acting on the system, for instance gravity.

Navier-Stokes equations can be considered as corrections to the Euler system for granular gases, just obtained from the set of equations (3)by dropping the terms involving λ, μ

and χ. This Euler system can be written in conservative form as

$$\begin{aligned}\frac{\partial \rho}{\partial t} + \operatorname{div}(\rho U) &= 0 \\ \frac{\partial(\rho U)}{\partial t} + \operatorname{div}\left[\rho\left(U \otimes U\right)\right] + \nabla p &= \rho F \\ \frac{\partial}{\partial t}\left(\frac{3}{2}\rho\theta + \frac{1}{2}\rho|U|^2\right) + \operatorname{div}\left[U\left(\frac{3}{2}\rho\theta + \frac{1}{2}\rho|U|^2 + p\right)\right] &= -\gamma + \rho U F.\end{aligned} \tag{4}$$

By analogy to classical molecular gases we will call $\epsilon = \frac{3}{2}\theta$ the enthalpy and thus, the total energy is given by $W = \frac{3}{2}\rho\theta + \frac{1}{2}\rho|U|^2 = \rho\epsilon + \frac{1}{2}\rho|U|^2$. The pressure is given by an equation of state of the form $p = p(\rho, \epsilon)$. The left-hand side of the equations corresponds to the conservation of number density, momentum and total energy.

Concerning the equation of state, it has been obtained [BCG00, Tos04] by means of expansions at small Knudsen and small inelasticity, that the first term in the equation of state is given by the standard pressure law in gas dynamics, i.e., $p = \frac{2}{3}\rho\epsilon$. However, corrections taking into account dense gas effects have been included in [JR85, GS95, BSSP04], the proposed corrected equation of state given by

$$p = \frac{2}{3}\rho\epsilon(1 + 2(1+e)G(\nu)). \tag{5}$$

Another phenomenological expression of the pressure has been proposed [HM03], in which the equation of state is

$$p = \frac{2}{3}\rho\epsilon + \alpha(\rho^2 - \rho_o^2)\mathcal{H}(\rho - \rho_o), \tag{6}$$

where α is a positive parameter, ρ_o is the close-packing density and $\mathcal{H}$ is an approximation to the unit step function. The correction term tries to mimic the incompressibility of the system at high densities.

Regarding the Navier-Stokes terms, we will write first the equations in conservative form, that is,

$$\begin{aligned}\frac{\partial \rho}{\partial t} + \operatorname{div}(\rho U) &= 0 \\ \frac{\partial(\rho U)}{\partial t} + \operatorname{div}\left[\rho\left(U \otimes U\right)\right] + \nabla p &= \operatorname{div}(P) + \rho F \\ \frac{\partial W}{\partial t} + \operatorname{div}\left[UW\right] &= -\operatorname{div}(q) + P : E + U \cdot \operatorname{div}(P) - \gamma + \rho U F,\end{aligned} \tag{7}$$

where the transport coefficients are taken from [JR85, BSSS99] for constant restitution coefficient. The bulk viscosity is given by

$$\lambda = \frac{8}{3\sqrt{\pi}}\rho\sigma\theta^{1/2}G(\nu)$$

and the shear viscosity by

$$\mu = \frac{\sqrt{\pi}}{6}\rho\sigma\theta^{1/2}\left[\frac{5}{16}\frac{1}{G(\nu)} + 1 + \frac{4}{5}\left(1 + \frac{12}{\pi}\right)G(\nu)\right],$$

the thermal conductivity is

$$\chi = \frac{15\sqrt{\pi}}{16}\rho\sigma\theta^{1/2}\left[\frac{5}{24}\frac{1}{G(\nu)} + 1 + \frac{6}{5}\left(1 + \frac{32}{9\pi}\right)G(\nu)\right]$$

and the cooling coefficient is

$$\gamma = \frac{12}{\sigma\sqrt{\pi}}(1 - e^2)\rho\theta^{3/2}G(\nu).$$

3 Numerical Method

The granular Euler equations (4) can be written as a very particular case of a system of nonlinear hyperbolic conservation laws with sources

$$\frac{\partial \mathbf{u}}{\partial t} + \mathrm{div}(\mathbf{f}(\mathbf{u})) = \mathbf{g}(\mathbf{u}) \tag{8}$$

for which there is an extensive literature in numerical methods.

We propose to use a finite difference characteristic-wise WENO method since it is a well-known shock-capturing and high-order method for nonlinear conservation laws. Extensive benchmarks have shown this method to be particularly well adapted to control oscillations in shock regimes in classical Euler equations for gases. We refer to the survey [Shu98] for all the details about WENO reconstruction procedures, benchmarks and references for different applications.

Let us remark that we can restrict ourselves to the 1D procedure, working in a dimension by dimension fashion for reconstructing each of the fluxes in (8). Given a uniform grid on the physical space, let us call $\mathbf{u}_i(t)$ the numerical approximation to the point value $\mathbf{u}(x_i, t)$. We construct a finite difference conservative (in the absence of sources) approximation to (8) by

$$\frac{d\mathbf{u}_i}{dt} = -\frac{\hat{\mathbf{f}}_{i+\frac{1}{2}} - \hat{\mathbf{f}}_{i-\frac{1}{2}}}{\Delta x} + \mathbf{g}(\mathbf{u}_i)$$

where $\hat{\mathbf{f}}_{i+\frac{1}{2}}$ is the numerical flux.

Let us briefly describe the general characteristic-wise finite difference WENO procedure with flux splitting. At each fixed time t, we find the numerical fluxes $\hat{\mathbf{f}}_{i+\frac{1}{2}}$ in the following way:

1. We first compute an average state $\mathbf{u}_{i+\frac{1}{2}}$ using Roe's formula [Shu98, KGSD00].
2. We compute the right and left eigenvectors and eigenvalues of the jacobian $\mathbf{f}'(\mathbf{u}_{i+\frac{1}{2}})$. These are explicitly known and formulas are postponed to the appendix.
3. We transform the values u_i and $\mathbf{f}(u_i)$ in the potential stencil of the approximation of the flux to the local characteristic coordinates by using the left eigenvectors.
4. We use Lax-Friedrichs flux splitting with the transport coefficient computed from the maximum, in absolute value, among the above computed eigenvalues.
5. We compute the decomposed fluxes at the middle points using WENO fifth-order reconstructions.
6. We transform back the computed fluxes to the physical space by using the right eigenvectors and add them to obtain $\hat{\mathbf{f}}_{i+\frac{1}{2}}$.

Let us remark that the previous procedure gives high order approximations of the convective part in the Euler system (4). Sources are treated explicitly and it seems from the numerical results that this treatment does not affect the overall performance of the solver.

We apply explicit 3rd-order TVD Runge-Kutta methods exposed in [SO88] to solve in time the resulting ODE system. As a consequence, we need to introduce a CFL condition, depending inversely on the maximum of the absolute values of the local eigenvalues computed in the second stage of the reconstruction process sketched above. In fact, it is easy to observe from explicit formulas in the appendix that this CFL condition becomes more and more restrictive in the close-packing limit since the sound speed (9) becomes infinite.

Navier-Stokes terms are approximated by centered second order finite differences and thus, additional conditions besides the CFL are imposed by taking the maximal values of the viscosity and thermal conductivity over the grid. It turns out that diffusion effectively paces the time stepping of the code. This fact is due to the dependences of the viscosity and thermal conductivity on the volume fraction, resulting in diverging coefficients for both close-packing and vacuum limits.

In the next two subsections we report the numerical results of two experiments: 1D sedimentation and 2D clustering as benchmarks for the solver.

3.1 1D Sedimentation

Here, we report the results of a simple sedimentation problem in which particles, already accelerated to Mach number 2, fall following gravity. Then, they collide with the $x = 0$ ground and they settle down. The spatial variable represents the particles height. Initial volume fraction and temperature are 0.02 and 1000 respectively.

In figure 1 we show in the left column the results of both Euler and Navier-Stokes systems upto the close-packing time in which maximal volume fraction is achieved at zero height. At this point the computation stops since the sound speed becomes infinity as $\nu \to \nu_{max}$.

By imposing the computation not to allow values of the volume fraction above 99% of the maximum volume fraction ν_{max}, then the computation can be continued. In the right

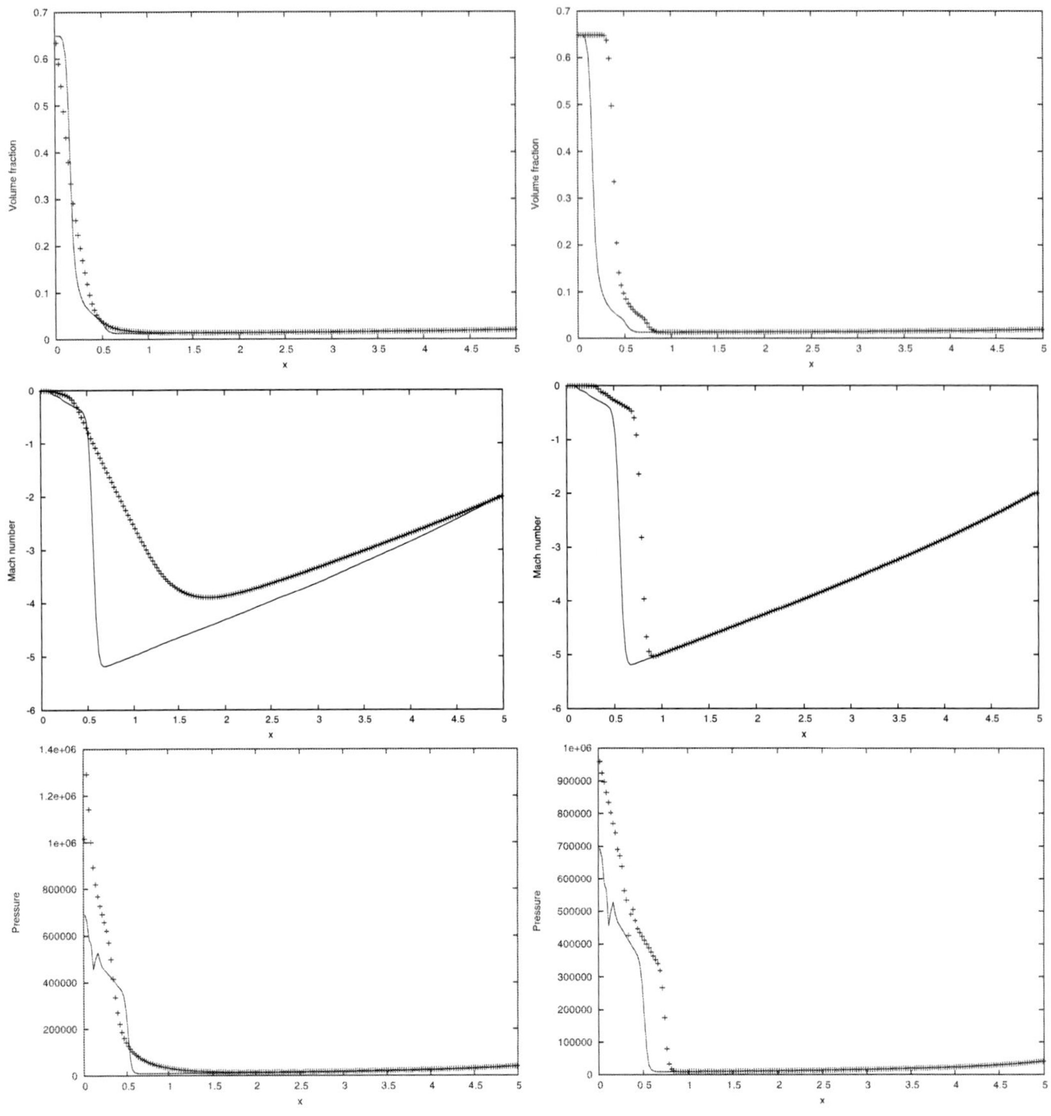

Figure 1: Left figures: comparisons between the results of Euler and Navier-Stokes systems till first close-packing time. Right figures: comparisons between the results of the Euler system for the first close-packing time and a later time with an imposed upper bound for the volume fraction.

column of figure 1, we compare the results of the Euler system upto the first close-packing time and the results of the continued computation with the imposed upper bound of the maximal volume fraction. We observe that the Mach number becomes zero and the volume fraction achives 99% of the value of ν_{max} on the ground.

3.2 2D Clustering

The interesting problem of cluster formation in a force-free granular gas constitutes a particularly difficult case for hydrodynamic modeling. In the classical numerical (molecular dynamics or MD) experiment reported first by Goldhirsch and Zanetti [GZ93], clusters arise from an initially homogeneous particle gas which cools down due to inelastic collisions. Thus, the system departs from the homogeneous cooling state characterized by the Haff law, for which the temperature, as a function of time, shows the well known dependence represented by a power law decay with the slope -2 at large times,

$$T(t) = \frac{T_0}{(1 - (1 - e^2) f_0 t)^2}$$

with f_0 being a constant related to the initial collision frequency, $f_0 = 4\sqrt{T_0/\pi} G(\nu)/\sigma$. The deviation from Haff's law reveals the appearance of particle aggregates or clusters, which grow progressively into bigger structures. In the late stages, the system with periodic boundaries shows a significant part of the material captured into a dense cluster, and vaste regions with no particles at all [LM98].

The difficulty of reproducing the system evolution via hydrodynamic simulations is clear, for the problematic limits ($\nu \to 0$, $\nu \to \nu_{max}$) will eventually be reached.

Rather than extracting interesting results from the simulated fields, here we are concerned with reproducing qualitatively the features of the cluster formation observed in MD simulations. For this purpose, we set an initially maxwellian particle gas equivalent to $N = 95475$ physical particles, distributed homogeneously inside a window over which a 200×200 grid is imposed. Periodic boundary conditions are set in both directions as in MD simulations. The restitution coefficient is chosen to be $\epsilon = 0.99$ and the global packing fraction $\nu_0 = 0.2$.

As expected, the evolution of the energy of the system reproduces the mentioned departure from Haff's law which underlines the development of the clustering phase (Fig. 2).

At time $\tau = (1 - e^2) f_0 t \sim 10$ in reduced units, inhomogeneities are in the order of 1% and the energy decay still follows Haff's law; at $\tau = 100$, inhomogeneities start to grow above 5% and the effect is already visible in the energy curve. In Fig. 3 we plot the contours of density at various representative stages, and Fig. 4 shows the corresponding Mach number.

Interestingly, a part from the overall increasing of the Mach number, Fig. 4 shows a brainlike structure spreading over the system. This structure, visible only until clusters develop, is clearly reminiscent of the velocity correlations associated to the vortex instability[BE98], the dark interfaces at low Mach numbers corresponding to the limiting contours between contiguous vortices.

The robustness of the code has made possible to reproduce cluster formation in a granular gas with constant restitution. However, long time simulations face the problem that, in

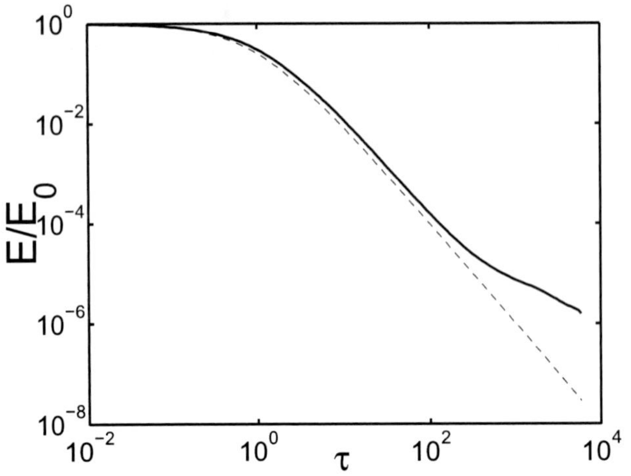

Figure 2: Solid line: total energy of the system over initial energy as a function of time in reduced units, $\tau = (1 - e^2) f_0 t$. Dashed line: Haff's law.

the course of the computation, temperature will eventually go to zero at some point close to a shock front. Due to the periodic boundaries though, the clusters are interacting with themselves much earlier as seen from the results, and therefore the usual question arises, whether further evolution of the system will provide meaningful results.

4 Appendix

Here, we summarize the formulas for the explicit eigenvalues and eigenvectors of the jacobian matrix of $\mathbf{f}(\mathbf{u})$ in one and two dimensions, for general equations of state depending on the density and the enthalpy (see also [KGSD00, SM04] for particular cases in which these formulas already were obtained). Let us remark that we keep the 3D volume fraction and the precise meaning of working in the 1D or 2D case is that we assume that the solution does not depend on the other spatial variables by symmetries of the problem.

In 1D, the vector $\mathbf{u} = (u_1, u_2, u_3)$ has components given by $u_1 = \rho$, $u_2 = \rho u$ and $u_3 = \rho\epsilon + \frac{1}{2}\rho u^2$; the corresponding flux functions are

$$f_1(\mathbf{u}) = u_2\,,\ f_2(\mathbf{u}) = p + \frac{u_2^2}{u_1}\,,\ f_3(\mathbf{u}) = \frac{u_2}{u_1}(p + u_3)$$

where the pressure is assumed to be a function of density and enthalpy,

$$p = p(\rho, \epsilon) = p\left(u_1, \frac{u_3}{u_1} - \frac{1}{2}\frac{u_2^2}{u_1^2}\right).$$

The sound speed is given by

$$c_s^2 = \frac{\partial p}{\partial \rho} + \frac{p}{\rho^2}\frac{\partial p}{\partial \epsilon} = p_\rho + \frac{p}{\rho^2} p_\epsilon, \tag{9}$$

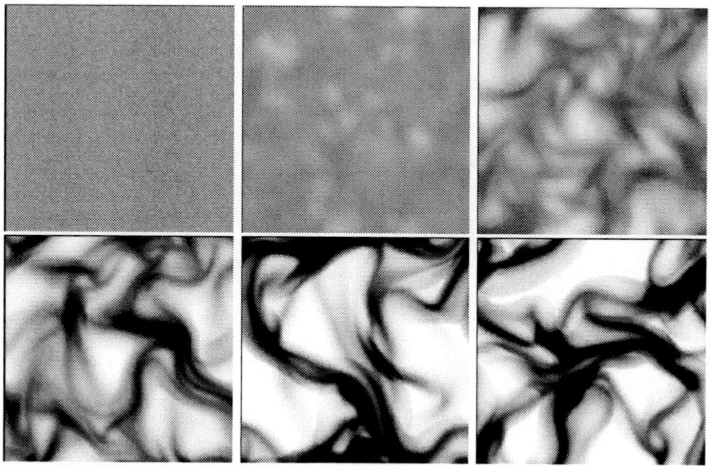

Figure 3: Density contour plots of the system at the following values of the reduced time $\tau = (1 - e^2) f_0 t$ (see text): 0, 0.9, 350, 850, 2000 and 6000.

by means of which, and along with the auxiliary function $H = \frac{p+u_3}{u_1}$, we can express the eigenvalues of the jacobian matrix $\mathbf{f}'(\mathbf{u})$ as

$$\lambda_- = \frac{u_2}{u_1} - c_s\,,\ \lambda_0 = \frac{u_2}{u_1}\,,\ \lambda_+ = \frac{u_2}{u_1} + c_s$$

and their corresponding right and left eigenvectors as

$$r_\pm \begin{pmatrix} 1 \\ \frac{u_2}{u_1} \pm c_s \\ H \pm \frac{u_2}{u_1} c_s \end{pmatrix} \qquad r_0 = \begin{pmatrix} 1 \\ \frac{u_2}{u_1} \\ H - \frac{1}{b_1} \end{pmatrix}$$

and

$$l_\pm \begin{pmatrix} \frac{b_2}{2} \mp \frac{u_2}{u_1}\frac{1}{2c_s} \\ -\frac{b_1}{2}\frac{u_2}{u_1} \pm \frac{1}{2c_s} \\ \frac{b_1}{2} \end{pmatrix} \qquad l_0 = \begin{pmatrix} \left[H - \frac{u_2^2}{u_1^2}\right] b_1 \\ \frac{u_2}{u_1} b_1 \\ -b_1 \end{pmatrix}$$

respectively. Here,

$$b_1 = \frac{p_\epsilon}{\rho c_s^2} \text{ and } b_2 = 1 + b_1 \frac{u_2^2}{u_1^2} - b_1 H.$$

In 2D, the vector $\mathbf{u} = (u_1, u_2, u_3, u_4)$ has components given by $u_1 = \rho$, $u_2 = \rho u$, $u_3 = \rho v$ and $u_4 = \rho\epsilon + \frac{1}{2}\rho|U|^2 = \rho\epsilon + \frac{1}{2}\rho(u^2 + v^2)$; and the corresponding flux functions of the derivative with respect to x are

$$f_{11}(\mathbf{u}) = u_2\,,\ f_{12}(\mathbf{u}) = p + \frac{u_2^2}{u_1}\,,\ f_{13}(\mathbf{u}) = \frac{u_2 u_3}{u_1}\,,\ f_{14}(\mathbf{u}) = \frac{u_2}{u_1}(p + u_4)$$

where the pressure and the sound speed are defined in the same way as before, and the auxiliary function is now $H = \frac{p+u_4}{u_1}$. The eigenvalues of the jacobian matrix $\mathbf{f}_1'(\mathbf{u})$ are given by

$$\lambda_- = \frac{u_2}{u_1} - c_s\,,\ \lambda_2 = \lambda_3 = \frac{u_2}{u_1}\,,\ \lambda_+ = \frac{u_2}{u_1} + c_s$$

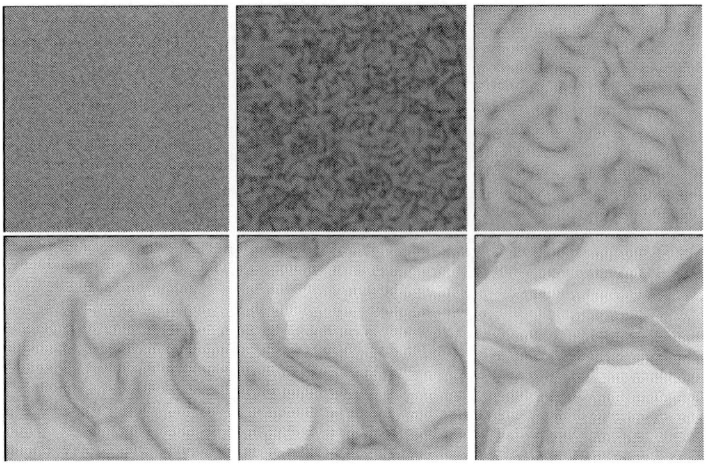

Figure 4: Mach number contour plots of the system at the same values of the reduced time as in Fig. 3: τ = 0, 0.9, 350, 850, 2000 and 6000. High values -white shades, low values -black shades.

and their corresponding right and left eigenvectors are

$$r_{\pm}\begin{pmatrix} 1 \\ \frac{u_2}{u_1} \pm c_s \\ \frac{u_3}{u_1} \\ H \pm \frac{u_2}{u_1} c_s \end{pmatrix} \quad r_2 = \begin{pmatrix} 1 \\ \frac{u_2}{u_1} \\ \frac{u_3}{u_1} \\ H - \frac{1}{b_1} \end{pmatrix} \quad r_3 = \begin{pmatrix} 0 \\ 0 \\ 1 \\ \frac{u_3}{u_1} \end{pmatrix}$$

and

$$l_{\pm}\begin{pmatrix} \frac{1}{2} \mp \frac{u_2}{u_1}\frac{1}{2c_s} + \frac{b_1}{2}\left(\frac{u_2^2+u_3^2}{u_1^2} - H\right) \\ -\frac{b_1}{2}\frac{u_2}{u_1} \pm \frac{1}{2c_s} \\ -\frac{b_1}{2}\frac{u_3}{u_1} \\ \frac{b_1}{2} \end{pmatrix} \quad l_2 = \begin{pmatrix} \left[H - \frac{u_2^2}{u_1^2} - \frac{u_3^2}{u_1^2}\right] b_1 \\ \frac{u_2}{u_1} b_1 \\ \frac{u_3}{u_1} b_1 \\ -b_1 \end{pmatrix} \quad l_3 = \begin{pmatrix} -\frac{u_3}{u_1} \\ 0 \\ 1 \\ 0 \end{pmatrix}$$

respectively. For the fluxes associated with the y-derivative, that is,

$$f_{21}(\mathbf{u}) = u_3\,,\ f_{22}(\mathbf{u}) = \frac{u_2 u_3}{u_1}\,,\ f_{23}(\mathbf{u}) = p + \frac{u_3^2}{u_1}\,,\ f_{24}(\mathbf{u}) = \frac{u_3}{u_1}(p + u_4)$$

we just need to swap indices 2 and 3 in all formulas above, and the second and third components in all right and left eigenvectors.

Acknowledgments

The first author acknowledges support from the European IHP network HYKE "Hyperbolic and Kinetic Equations: Asymptotics, Numerics, Applications" HPRN-CT-2002-00282, the DGI-MCYT/FEDER project BFM2002-01710 and CIRIT 2001SGR/00172. The second author acknowledges support from the DGI-MCYT/FEDER project DPI2003-06725-C02-01 and CIRIT 2002SGR00189.

References

[BSSS99] Bizon, C., Shattuck, M.D., Swift, J.B., Swinney, H.L.: Transport coefficients for granular media from molecular dynamics simulations. *Phys. Rev. E*, **60**, 4340–4351 (1999)

[BCG00] Bobylev, A.V., Carrillo, J.A, Gamba, I.: On some properties of kinetic and hydrodynamics equations for inelastic interactions. *J. Statist. Phys.*, **98**, 743–773 (2000)

[BMSS02] Bougie, J., Moon, Sung Joon, Swift, J.B., Swinney, H.L.: Shocks in vertically oscillated granular layers. *Phys. Rev. E*, **66**, 051301-1 to 051301-9 (2002)

[BDKS98] Brey, J.J., Dufty, J.W., Kim, C.S., Santos, A.: Hydrodynamics for granular flows at low density. *Phys. Rev. E*, **58**, 4638–4653 (1998)

[BSSP04] Brilliantov, N.V., Salueña, C., Schwager, T., Pöschel, T.: Transient structures in a granular gas. *Phys. Rev. Lett.*, **93**, 134301-1 to 134301-4 (2004)

[BP100] Brilliantov, N.V., Pöschel, T.: Granular gases with impact-velocity dependent restitution coefficient. In: Pöschel, T., Luding, S. (eds) *Granular Gases. Lecture Notes in Physics*, **564**. Springer Verlag, Berlin, 100–124 (2000)

[BP200] Brilliantov, N.V., Pöschel, T.: Granular Gases - The early stage. In: Rubi, M. (ed) *Coherent Structures in Classical Systems*. Springer, (2000)

[BE98] Brito, R., Ernst, M.H.: Extension of Haff's cooling law in granular flows. *Europhys. Lett.*, **43**, 497–501 (1998)

[CIP94] Cercignani, C., Illner, R., Pulvirenti, M.: The mathematical theory of dilute gases. *Springer Series in Applied Mathematical Sciences*, **106**. Springer-Verlag, (1994)

[DLK95] Du, Y., Li, H., Kadanoff, L.P.: Breakdown of hydrodynamics in a one–dimensional system of inelastic particles. *Phys. Rev. Lett.*, **74**, 1268–1271 (1995)

[GZ93] Goldhirsch, I., Zanetti, G.: Clustering instability in dissipative gases. *Phys. Rev. Lett.*, **70**, 1619–1622 (1993)

[GSBMSS98] Goldman, D., Shattuck, M.D., Bizon, C., McCormick, W.D., Swift, J.B., Swinney, H.L.: Absence of inelastic collapse in a realistic three ball model. *Phys. Rev. E*, **57**, 4831–4833 (1998)

[GS95] Goldshtein, A., Shapiro, M.: Mechanics of collisional motion of granular materials. Part 1. General hydrodynamic equations. *J. Fluid Mech.*, **282**, 75–114 (1995)

[KGSD00] Kamenetsky, V., Goldshtein, A., Shapiro, M., Degani, D.: Evolution of a shock wave in a granular gas. *Phys. Fluids*, **12**, no. 11, 3036–3049 (2000)

[Haf83] Haff, P.K.: Grain flow as a fluid-mechanical phenomenon. *J. Fluid Mech.*, **134**, 401–430 (1983)

[HM03] Hill, S.A., Mazenko, G.F.: Granular clustering in a hydrodynamic simulation. *Phys. Rev. E*, **67**, 061302 (2003)

[JR85] Jenkins, J.T., Richman, M.W.: Grad's 13-moment system for a dense gas of inelastic spheres. *Arch. Rational Mech. Anal.*, **87**, 355–377 (1985)

[JS96] Jiang, G., Shu, C.-W.: Efficient implementation of weighted ENO schemes. *J. Comput. Phys.*, **126**, 202–228 (1996)

[LM98] Luding, S., McNamara, S.: How to handle the inelastic collapse of a dissipative hard-sphere gas with the TC model. *Granular Matter*, **1**, 113–127 (1998)

[LSJC84] Lun, C.K.K., Savage, C.B., Jeffrey, D.J., Chepurniy, N.: Kinetic theories of granular flow: inelastic particles in Couette flow and slightly inelastic particles in a general flow field. *J. Fluid Mech.*, **140**, 223–256 (1984)

[SM04] Serna, S., Marquina, A.: Capturing Shock Waves in Inelastic Granular Gases. *CAM report* 04-04.

[RPBS99] Ramírez, R., Pöschel, T., Brilliantov, N.V., Schwager, T.: Coefficient of restitution of colliding viscoelastic spheres. *Phys. Rev. E*, **60**, 4465–4472 (1999)

[RBSS02] Rericha, E., Bizon, C., Shattuck, M., Swinney, H.L.: Shocks in Supersonic Sand. *Phys. Rev. Lett.*, **88**, 1–4 (2002)

[SG98] Sela, N., Goldhirsch, I.: Hydrodynamic equations for rapid flows of smooth inelastic spheres, to Burnett order. *J. Fluid Mech.*, **361**, 41–74 (1998)

[Shu98] Shu, C.-W.: Essentially non-oscillatory and weighted essentially non-oscillatory schemes for hyperbolic conservation laws. In: Cockburn, B., Johnson, C., Shu, C.-W., Tadmor, E. (Quarteroni, A., ed) *Advanced Numerical Approximation of Nonlinear Hyperbolic Equations. Lecture Notes in Mathematics*, **1697**. Springer, 325–432 (1998)

[SO88] Shu, C.-W., Osher, S.: Efficient implementation of essentially non-oscillatory shock capturing schemes. *J. Comput. Phys.*, **77**, 439–471 (1988)

[SR04] Swinney, H.L., Rericha, E.C.: Pattern formation and shock waves in granular gases. In: Mallamace, F., Stanley, E. (eds) *The Physics of Complex Systems (New Advances and Perspectives)*. The International School of Physics Enrico Fermi, Italian Physical Society, (2004)

[Tos04] Toscani, G.: Kinetic and hydrodynamic models of nearly elastic granular flows. *Monatsh. Math.* **142** 179-192 (2004)

INDEX

Q

R

S

T